EXTINCTIONS

MICHAEL J. BENTON

EXTINCTIONS

How Life Survives, Adapts and Evolves

Frontispiece: Stromatolites on early Earth. These structures, built from layers of algae and mud, represent some of the earliest life we know.

Endpapers: Detail of lava during a volcanic eruption. Budkov Denis/ Shutterstock

First published in the United Kingdom in 2023 by
Thames & Hudson Ltd, 181A High Holborn, London WC1V 7QX

First published in the United States of America in 2023 by
Thames & Hudson Inc., 500 Fifth Avenue, New York, New York 10110

Extinctions © 2023 Thames & Hudson Ltd, London
Text © 2023 Michael J. Benton
Designed by Matthew Young

British Library Cataloguing-in-Publication Data
A catalogue record for this book is available from the British Library

Library of Congress Control Number 2023935253

ISBN 978-0-500-02546-8

Printed in China by Shenzhen Reliance Printing Co. Ltd

FSC
www.fsc.org

MIX
Papier aus verantwor-
tungsvollen Quellen
FSC® C102842

Be the first to know about our new releases,
exclusive content and author events by visiting
thamesandhudson.com
thamesandhudsonusa.com
thamesandhudson.com.au

Contents

Preface

The great conservationist E. O. Wilson called the love of nature, and especially the love of biodiversity, biophilia. He had in mind not only our love of the outdoors, of nature, of getting away from our hurried, technological world, but also a fundamental justification for wildlife conservation. Why should we prevent species from going extinct? We could make economic arguments and say that we get useful products such as food and drugs from many plants, or that natural habitats keep the Earth's energy flows, oxygen and carbon in balance. But Wilson was keen to remind us that purely economic arguments are not enough, or they can be used misleadingly. His point was that we should love biodiversity, the richness and colour of all living things, and humans (or any other species) have no right to kill off all members of another species.

But then, as a lad of seven, when I was introduced to dinosaurs, and fossils in general, I loved the fact they were extinct. I could imagine these past worlds of trilobites, great bone-armoured fishes, ichthyosaurs, dinosaurs and mammoths. How different they must have been! Who needs science fiction imaginings when the real thing, the world in front of us, has passed through such enormous spans of time and experienced so many alien worlds? And the evidence of these thousands upon thousands of extinct species is there in the rocks, preserved as fossils. I dreamt then, as I looked at my dinosaur books, that someday I could dig up these fossils and bring them back to life, not literally, but by using all the smart tools in the scientific laboratory to work out whether a particular dinosaur was warm-blooded or not, whether a giant pterosaur could fly, and how a 50-tonne sauropod could find enough food to keep its huge body functioning successfully. And we care about more than dinosaurs; we wonder at creatures like *Hallucigenia* and *Anomalocaris*, which represent some of the earliest marine animals of the Cambrian explosion. Even their names tell us they are weird and amazing, anomalies, or visions of a bad dream. We read about the topsy-turvy ecosystems that existed on Earth after the dinosaurs disappeared, and a time 60 million years ago when giant land-walking crocodiles were the main predators in some places, and giant, flightless, horse-eating birds terrorized life in South America.

This is 'palaeobiophilia', a mouthful of a word I have invented for this viewpoint, meaning 'the love of ancient life for its own sake'. There are no arguments here that the fossils should be in some way useful or carry a moral lesson for us. They don't have to be closely related to living species or not, or show some amazing property such as being the largest this or the oldest that. It's enough that they really existed, and we can know something about them and wonder at their stories. Each fossil species had a beginning and an end. The end is the extinction of the species, the point at which the last population or individual died, the snuffing out of all the history to that point, which was preserved in the genetic code of the last individuals; for each species there was a reason why it finally disappeared.

Extinctions happen on different scales. What we usually think about are extinctions of single species, such as the complete annihilation of the dodo or quagga. Each such extinction has a cause, and in these two cases, it could presumably be laid at the door of a single person whose name has by now been forgotten. Single species extinctions, however, have happened ever since life first evolved. No species lasts forever; in fact, species of mammals and birds typically last for about a million years, and for some other groups, such as molluscs and some plants, individual species

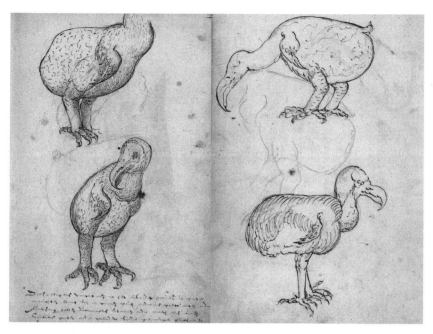

Is this what the dodo really looked like? Sketches from the ship's journal of the *Gelderland*, 1601, showing living and recently killed specimens.

might last for ten million years. These are long spans of time, but when we recall that the Earth is over 4.5 billion years old, species are short-lived. They come and go. Some die out because food runs out in their local patch, or the climate becomes too warm or too cold. Others die out because a vigorous incomer steals all their food or space. These are what palaeontologists call 'background extinctions'. Although this is a rather dismissive term, it forms a comparison with more widespread incidents, typically called 'extinction events'.

An extinction event is when many species die out at the same time, perhaps for related reasons. They can be regional, such as when the Mediterranean Sea dried up during the Messinian Salinity Crisis from 5.96 to 5.33 million years ago. What had been sea turned into land when the western connection of the Mediterranean to the Atlantic, located between Spain and Gibraltar in the north and Morocco in the south, closed off. The trapped seawater evaporated, leaving thick salt deposits, hence the name 'salinity crisis'.

All the fishes, shellfish and other marine life died. Many of these were local populations of widespread species and so the whole species did not go extinct. But some species of fishes and molluscs were restricted to the Mediterranean and so some went extinct for the same reason and at the same time.

Extinction events may only affect particular kinds of species, as happened at the end of the Pleistocene ten thousand years ago, when the last of the northern hemisphere ice sheets retreated from North America, northern Europe and Asia. Cold-adapted species such as mammoths, mastodons, woolly rhinoceros and cave bears headed north with the retreating ice, but eventually ran out of space and food and died out, although some of these species were probably helped on their way by early human hunting bands.

Some extinction events were much larger, and so huge they are called 'mass extinctions'. These were times when thousands or millions of species died out at the same time and in all parts of the world, representing a broad array of plants and animals living in the sea and on land. Humans have never witnessed a mass extinction, although our ancestors in all parts of the world did witness the extinctions of large mammals ten thousand years ago. Palaeontologists have identified five mass extinctions in the geological record: at the end of the Ordovician, in the Late Devonian and at the ends of the Permian, Triassic and Cretaceous geological periods, respectively 444, 372 (and 359), 252, 201 and 66 million years ago. These are the 'big five', and the reason why many commentators have identified the current biodiversity crisis as the 'sixth mass extinction'.

How are we to look at these extinctions, especially in the context of deep, geological time? If we weep over the death of the dodo, should we also lament the thousands or millions of species that died out in each of the mass extinctions of the past? In the most famous of these at the end of the Cretaceous, not only did dozens of species of dinosaurs disappear, but also the pterosaurs (great flying reptiles) and various marine reptile groups. At the same

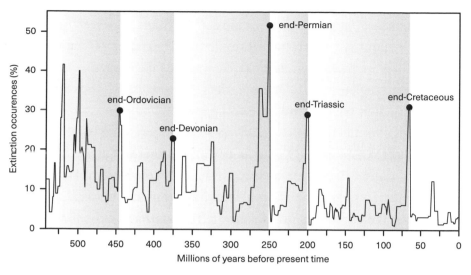

The 'big five' mass extinctions, identified as times of especially high extinction rates through the last 540 million years.

time, many birds and mammals died out on land, as did the abundant swimming molluscs, ammonites and belemnites, as well as a great number of planktonic species. These species would mainly have survived, at least for a short time geologically speaking, if a meteorite had not hit the Earth.

But mass extinctions also have a creative aspect. Life has always bounced back. After the killing agency has gone – whether a meteorite or a huge volcanic eruption – the surviving species take over. They occupy very strange ecosystems for a while, which have great holes. An ecosystem is the sum of species and their interactions associated with their particular physical habitat in a particular place. A mass extinction might knock out half of the species in an ecosystem. Taking as an example a modern woodland in North America, the rabbit might survive, but some of its predators, perhaps coyotes and foxes, disappear. The shrews and mice might still be there snuffling in the undergrowth, but the voles might have gone, as, too, some of their predators like owls and sparrow hawks. The survivors might also be suffering from shock, and even show injuries from their traumas. But they go searching for food, and life goes on. After a few generations, numbers build up and damaged landscapes recover. After a few thousand generations, the survivors might have changed in noticeable ways, changing size or diet, taking advantage of new opportunities. New species emerge, the food webs build back and eventually a new ecosystem has become established.

What is often also seen in these times of recovery from devastation is something completely new, the first swimming scallop or the first dinosaur, for example. Something special can happen during the rebuilding of ecosystems when the removal of dominant plants and animals gives others their opportunity to take over. It seems that there is a kind of inertia in ecosystems, where the same stable system goes on and on because the effort, in terms of ecology or evolution, is too much to shake up the old regime. This is what ecologists Michael Rosenzweig and Robert McCord called 'incumbent advantage'. If you are part of the ruling system, you are protected simply by being there.

The best example of incumbent advantage and replacement in the fossil record is probably the case of dinosaurs and mammals 66 million years ago. The story had always been that the mammals succeeded because they were, and are, smart, warm-blooded and cared for their babies. The dinosaurs somehow deserved to die because they were huge, dim-witted, cold-blooded and didn't care for their young at all. When I started my career as a palaeontologist, it was then fashionable to reject such moral tales. Just because we are mammals doesn't mean mammals are best. In fact, reptiles have their place, and their slower metabolism and water conservation adaptations mean they can thrive in hot and dry places. And, of course, the facts themselves didn't quite fit; so far as we could tell, dinosaurs were warm-blooded, too, so mammals didn't necessarily have that advantage. And indeed, mammals evolved in the Late Triassic at about the same time as the first dinosaurs, so if they were so smart, why didn't they take over the world then? In the new world of palaeontology, we demanded proof, not moral tales.

It turns out the answer was the incumbent advantage of dinosaurs. In numerical analyses carried out in 2013, Graham Slater, a palaeobiologist then at the Smithsonian Museum in Washington, D.C., explored all possible models for the evolution of dinosaurs and early mammals, including models where they effectively did not interact at all, where everything was driven by the end-Cretaceous mass extinction and various models of competitive interaction. The winning model was 'release and radiate', meaning mammals had been held back by the incumbent dinosaurs and could only freely diversify after the dinosaurs had been removed from the scene. The mass extinction gave them their great opportunity, and here we are today. The removal of dinosaurs and their continual trampling of plants probably also gave the flowering plants a chance to explore their potential: the first of the modern-type tropical rainforests with their enormous biodiversity also date back to that post-extinction time of renewal.

So, we can see that mass extinctions can be creative when we look at the long span of the history of life. Other examples are presented in the book where, for example, mass extinctions at the end of the Permian and in the Triassic triggered or enabled the evolution of modern-style ecosystems in the sea and on land – we trace such familiar phenomena as coral reefs, fast-swimming fishes, modern sharks and molluscs in the oceans, as well as flies and beetles, modern conifer trees, frogs, lizards, crocodiles and mammals on land, back to the opportunities for new evolution that had been generated by the mass clear-out of incumbent species through mass extinction.

Is extinction a good or bad thing? Surely, it's always bad? We must certainly regret that our actions can cause the loss of particular species forever: we killed the dodo and will never get to see this lovely bird. Its place on the island of Mauritius is empty and nothing has replaced it, so in the balance of nature, there is a place for the dodo. The dodo was a remarkable and unique creature as witnessed by the sailors of many nations before the last ones were wiped out – maybe not simply to provide a reportedly slightly unpleasant feast. In fact, as we will see, dodos might still be here if that was the only reason they died. It was more likely that the rats and cats humans introduced to Mauritius did for them.

But that does not exculpate humans: whether we kill them and eat them, kill them and don't eat them, or simply perturb their ecosystems by our disruptive behaviours, spreading disease, rats, cats, sparrows and rabbits, humans have a habit of spoiling nature. We have caused the extinction of countless documented species, like the dodo; we have doubtless also killed off many other species we never knew about. As Edward O. Wilson stressed in his many books about biodiversity and extinction, humans have a responsibility to preserve all of life in its rich diversity, and this is a message emphasized by David Attenborough in countless television and radio programmes, and now a core political message of Greenpeace, Friends of the Earth and especially Extinction Rebellion.

Is the answer, then, to bring extinct species back to life? And if we could, would that be a good thing? There is indeed a very active movement to bring species from the past back to life. We saw it from the 1990s onwards in the *Jurassic Park* movies and in travellers' tales of exquisitely frozen mammoth carcasses, so perfect you could even eat the flesh. The *Jurassic Park* scenario, presented first by Michael Crichton in his 1990 book, was that a sample of blood from a dinosaur could be extracted from a mosquito preserved in amber, cloned to make a larger sample and injected into a frog, thus eventually producing a blueprint for a dinosaur. At the time, early dinosaur DNA analysts thought this, in fact, might be possible. Since then, we have realized that DNA does not survive well at all and it is even difficult to get usable fragments of DNA from recently extinct animals such as the quagga, a zebra-like animal that was driven to extinction in South Africa in the late 1800s. Early reports of Mesozoic-aged DNA from plants and insects, and even from dinosaurs, all turned out to be contamination in the lab.

So, we can't bring dinosaurs back to life in this way. But what about unfreezing frozen Ice Age mammals? Until recently the realm

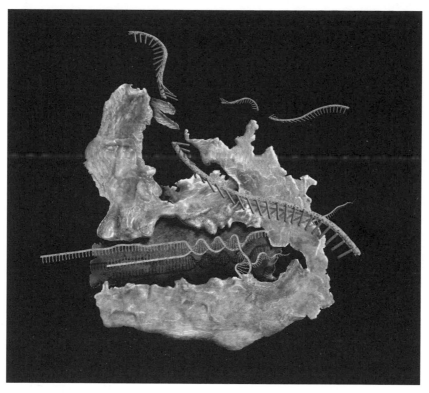

Protein synthesis in the cell. The double-stranded DNA helix (lower centre) provides the instructions for messenger RNAs (top) to gather amino acids to manufacture proteins in the ribosome (blobby background structure).

of eccentric scientists and mavericks, genetic engineering now makes such a dream feasible. For example, the intention of the Colossal Company, based in Texas in the USA and backed with millions of dollars of funding, is to bring the mammoth back from the dead. As the company says on its website, 'Extinction is a colossal problem facing the world ... and Colossal is the company that's going to fix it.' The de-extinction of the mammoth will be achieved by genetically engineering the germ-line information in the DNA of a modern elephant to insert genes that enable them to survive in colder conditions than normal. Farmers have long applied selective breeding techniques to make their pigs fatter and their corn crops richer. By adding genetic engineering to their armoury, crops can become even richer or can survive in colder or hotter conditions than normal.

The same approaches could make Asian elephants adapted to live in cold conditions so that they might be induced to live in Siberia and could enable the tundra to flourish again as it did when they were grazing across northern Russia and Canada.

Is there room in modern ecosystems for mammoths? Perhaps. Dodos, yes perhaps also. But what about the other extinct elephants, horses and giant sloths that populated North America until ten thousand years ago? Humans have squeezed the space for wild nature so much in the past two hundred years that there isn't enough room for the ten million species on Earth today, let alone reintroducing every extinct species as well. In fact, living species probably represent less than 1% of all that have ever existed on the Earth, and who are we to judge which de-extinct species could or could not find a place without massively perturbing the existence of those already there? Our introductions of non-native species have already caused more extinctions of native species than many other acts of human foolishness – for example, when British explorers transported sparrows and rabbits to Australia without any regard for the kangaroos and other marsupials already living there.

If extinction is always bad, then de-extinction ought to be seen as broadly good. But, as we have seen, the argument is much more subtle and each case has to be qualified with 'well, it depends...'.

But is it always a tragedy when a species becomes extinct? I am going to argue a rather opposite view in this book. Palaeontology shows us that many billions of species that once existed are now extinct, and their natural extinctions enabled new species to inherit the Earth. However, my case is certainly not in favour of humans causing the extinction of any species. There is no doubt that each such extinction is a tragedy, evidence of a terrible, arrogant and foolish series of actions by selfish people who in many cases were warned of the consequences of their actions in destroying a patch of forest or using particularly fiendish fishing technology. How do we make a case *for* nature?

PART 1

Origins

4,567–444 million years ago

CHAPTER 1

The First Animals and Mass Extinctions

THE GARDEN OF EDIACARA

The world more than 555 million years ago was very different from today (see Plate III). For a start, there was no life on land, or at most some simple seaweeds and microscopic organisms that clung to the edges of the oceans. But on the seabed in many parts of the world, we enter the strange world of the Ediacaran animals. They have been known only since 1947 but it has been hard to resolve what they are. Do they include examples of early sponges, corals, worms and starfish … or are they something else altogether?

The Ediacaran seabed was a lively place. As you dive below the surface, you first see tall fleshy blades, waving sinuously in the gentle seabed current, each shaped like a fern frond, flat and with deep grooves marked on the front and back. The fronds are up to 30 cm (1 ft) long, and on close inspection you can see that they are all divided into smaller branches, something like the complex make-up of a feather. These are *Charnia*, one of the most famous of the Ediacaran fossils, and they are abundant in the assemblage, standing as close together as trees in a forest, each one fixed to the seabed with a root-like holdfast structure.

Lying flat on the sea floor between the *Charnia* fronds is a tiny disc-shaped organism that looks like a Catherine wheel firework rotating around a pin, spitting sparks. This animal is circular, only 10 mm (½ in.) across. It has grooved structures on top that radiate spirally from the centre, forming three segments or branches. *Tribrachidium*, named for this 'three-branched' pattern of symmetry, seems to spend most of its time immobile, sitting on the seabed and filtering small particles of food from the seawater that flows by. Experiments with three-dimensional models in flow tanks have shown that the passing water was funnelled by the three arms and slowed down over the creature's mouth-like pits, perhaps allowing food particles to drop out.

Moving deeper into the frond jungle, you spot a broad, flat animal, somewhat circular in outline and shaped like a flatfish, but with lateral segment grooves running from a midline to the edges. It moves slowly over the sea floor, flopping over several *Tribrachidium* and stirring up the sand. The largest creature of its day, *Dickinsonia*

Fossils of two of the most famous Ediacaran animals, the frond-like *Charnia* (left) and the disc-like, segmented *Dickinsonia*. Both are hand length.

fossil specimens range in size from a few millimetres to 1.4 m (4 ft 7 in.) in length. There are several species, some of them nearly circular in outline, others elongate and shaped like the glass in divers' goggles. There are even babies and adults, the babies like small coins, growing to full size by the addition of more and more segments. *Dickinsonia* may have fed by chomping bacterial growths on undisturbed areas of the sea floor using means that remain mysterious – whether it had some sort of mouth at the front end, just tucked underneath the body disc, and slurped the food in, or used some chemical means to dissolve organic matter and somehow absorb it through its fleshy underside cannot yet be said.

Crawling along unperturbed by the antics of the *Dickinsonia* are a couple of segmented, 3–5-cm (1–2-in.)-long animals called *Spriggina* that look like elongate woodlice, with their heads covered by a single round-fronted shield. They twist and turn from side to side, but do not seem to have any legs under their carapace. They might have pressed against holdfasts of *Charnia* to make turns, or perhaps had some fleshy structures underneath that enabled them to swim and these have not been preserved in the fossils.

Another delicate animal twists and turns in slow motion on the seabed, moving along like a spaced-out snail or limpet, hoovering up particles of organic matter. As it goes, it leaves loosely coiling traces of its passage in the algal mat on the seabed. This is *Kimberella*, a snail-like animal with an oval body shape and a pair of short prongs extending out at the front. Behind the head, the body is built up as an oval structure above the flat base, with short ridges radiating to the edges of the base plate. This frilled fringe might have been involved in respiration, a means to extract oxygen from the seawater, but that is conjecture. Over a thousand specimens of *Kimberella* have been found at the famous White Sea area in Siberia, as well as in Australia and Iran, and they range in length from 1 to 15 cm (½–6 in.).

There are all sorts of other strange frond-like and worm-like creatures in the Ediacaran garden. They are commonly reconstructed by palaeoartists in reds, blues, greens and pinks, but these colours are entirely imaginary. For all we know, the Ediacaran beasts might have been dull brown or grey in colour.

FINDING FIRST LIFE: A PARADOX

The Ediacaran animals were first reported in 1947 by Reg Sprigg (1919–1994), an Australian state geologist who was searching for economically important minerals in the Ediacara Hills in South Australia. The hills are located 650 km (400 miles) north of Adelaide and form part of the Flinders Ranges, the largest series of mountains in Australia. The Ediacaran landscape is blisteringly hot and dry in summer, with extensive exposure of the red-brown sandstones. Sprigg knew that the rocks were Precambrian in age, meaning they were hugely ancient, representing some part of the vast span of time from the origin of the Earth 4,567 million years ago to the Cambrian Period, which began 540 million years ago.

Geologists are practical folk: the already-named Cambrian marked the stratigraphic range of rocks in which fossils could be found, which was then believed to match the time period when all groups of animals originated; when rocks older than the Cambrian were found, they were appropriately assigned to the Precambrian. The frustration is that these earliest of all rocks span such a vast amount of time – in fact, fully 88% of the history of the Earth – and yet so little is known about this great eon.

Sprigg certainly had no expectation of finding any fossils, let alone identifiable fossils of animals, in the Ediacaran rocks. If he had read Charles Darwin, he might have recalled that the great father of evolution had written in 1859 'that during these vast, yet quite unknown, periods of time, the world swarmed with living creatures. To the question why we do not find records of these vast primordial periods, I can give no satisfactory answer.' Darwin (1809–1882) could not explain why Precambrian fossils had not been found, but he was sure that eventually they would be discovered and provide evidence of the earliest stages in the evolution of life. The problem, or even the paradox, is that although we would dearly love to know everything about the very earliest steps in the origins of life, by definition the earliest forms of life would be simple, so fossils might be hard to determine for this reason.

How simple? Bacteria are some of the simplest forms of life today, each organism consisting of a single cell. They don't even have a nucleus, the structure at the core of each of our cells that contains the DNA – the genetic code that not only enables individual cells to multiply and replace themselves correctly, but also to reproduce and pass on genetic information to the next generation to make sure their bodies form correctly. Single-celled bacteria have very basic DNA, no cell nucleus and otherwise simplified cell systems. They are also tiny, about 2 micrometres across – that is, 2-millionths of a metre.

Added to the expected simplicity of earliest life, Precambrian rocks are sparse and modified by virtue of their great age. The very oldest rocks have often been buried, with high pressure and high temperature changing sedimentary rocks such as sandstones and mudstones into metamorphic rocks such as quartzites and slates. Any fossils, whether large or small, would tend to be squashed, bent or even obliterated.

So, despite Darwin's hopeful prediction of oceans swarming with life in the Precambrian, practical geologists were not so sure. Indeed, in the 1830s, as a young man, Darwin had tramped over the slate hills of North Wales with his geologist friend Adam Sedgwick from the University of Cambridge and seen typical Cambrian fossils such as trilobites (see Chapter 2), but even these Cambrian fossils were rare and sometimes hard to make out in the metamorphosed slate rocks. Geologists since Darwin's day did not expect to find much in the earliest rocks, and nor did Reg Sprigg as he tramped the Flinders Ranges, mapping out the geology.

PRECAMBRIAN LIFE, OR NOT?

If you don't have any fossils, you can't identify any extinctions, and not much had been identified from the Precambrian rocks from the 1830s up to 1947. From time to time, there was a flurry of excitement when fossils were announced, and palaeontologists are nothing if not optimists. But nearly all these trumpeted finds were rejected for one reason or another; they were either inorganic blobs in the rock formed by chemical means, marks in the sediment made by raindrops or bubbles, or contamination by modern fluff or microbes on the microscope slides and not Precambrian in age at all. So palaeontologists in the first half of the twentieth century were just as much in the dark about the origins of major living groups such as microbes, plants and animals as they were in Darwin's day.

The only widely accepted evidence to support Darwin's expectation that fossils would be found in the Precambrian were stromatolites, meaning 'layered rocks'. Modern stromatolites were not identified as biological in origin until the 1950s; examples are known from shallow, warm oceans such as those along the west coast of Australia, famously at Shark Bay. There, stromatolites build up from bacterial films in shallow waters. A thin layer of green slime forms as the microscopic cells split. Then, a flurry of currents may shoot sand grains and mud over the living mat. The single-celled organisms either move up or send thin tendrils of living tissue upwards, and these form a new layer of living mat. The organisms are called cyanobacteria, and they are unusual in that they photosynthesize: they are not plants, but like plants, they contain the green chemical chlorophyll that helps transform water, carbon dioxide and minerals into oxygen and organic compounds to build their bodies, using the energy from sunlight. Green plants and green cyanobacteria grow in places where they can capture sunlight, and so they stay in shallow water or on land. Over time, the interleaved layers of cyanobacterial mat and mud

The famous stromatolites from Shark Bay in northwestern Australia, modern examples of some of the oldest known biological structures on Earth, produced by layers of microscopic, photosynthesizing cyanobacteria and mud.

can build into quite a stack, sometimes many metres thick. Many of the Precambrian stromatolites were built by cyanobacteria, but the oldest ones occur before this group originated and were formed by specialist microbes that could photosynthesize in the absence of oxygen.

The great thing for palaeontologists is that stromatolites are not only a sure indication of life, but also large and visible to the eye, so can be recognized in the fossil record; geologists had already identified them in Cambrian and younger rocks. In 1883, Charles Walcott (1850–1927), an American palaeontologist who worked for the United States Geological Survey, was sent out to explore the American Midwest. He and his team travelled down the Colorado River on rowing boats and hand-made rafts, and as they hurtled and jostled through the racing waters in the Grand Canyon, in the banks of the river they identified the first Precambrian stromatolites ever seen – and the first solid evidence of Precambrian life. These were overall circular structures, looking something like squashed cabbages, showing the characteristic repeated interlayered structure of mud and organic matter. Later, Walcott described a large, circular alga, measuring 5 mm (¼ in.) across, from the same rocks: the first acceptable evidence of an individual, identifiable Precambrian fossil. These finds were from rocks assigned to the so-called Chuar Group, now known to be about 750 million years old, part of the final major division of Precambrian time called the Neoproterozoic.

Despite the importance of the finds and Walcott's ability (he also discovered the Burgess Shale fossils; see page 37), his Precambrian fossils were rejected at the time by the doyen of palaeobotany, Albert Charles Seward (1863–1941), Professor of Palaeobotany at the University of Cambridge in the UK. In 1931, Seward wrote that Walcott's claims were, 'I venture to think, not justified by the facts. It is clearly impossible to maintain that such bodies are attributable to algal activity … we can hardly expect to find in Pre-Cambrian rocks any actual proof of the existence of bacteria.' Walcott could not defend his case as he had died four years earlier. By 1940, the general consensus remained that there were no fossils to be found in Precambrian rocks.

PIECING TOGETHER THE STORY

Everything changed after Reg Sprigg's discovery of the first Ediacaran fossils in 1947. Their age was debated, but they were accepted as fossils. In 1957, the first examples of *Charnia* were reported from England, and since then, Ediacaran fossils have been recognized from Russia, Canada, Namibia and many other locations, spanning from 602 to nearly 540 million years ago. In 1953, Stanley Tyler, a geologist from the University of Wisconsin, reported simple single-celled fossils from the Gunflint Chert of

Ontario, Canada. Chert is a glassy rock made from silica that can preserve fine-scale organic fossils. Tyler's Gunflint fossils were immensely old at 1,900 million years, and included both single cells and filaments made from many cells. Then, in 1956, the Shark Bay stromatolites were discovered, and the work of Walcott and many others was vindicated. Modern examples proved that stromatolites were indeed made from living cyanobacterial or microbial mats and critics such as Seward had been wrong. Precambrian studies took off after the 1950s because they had become fruitful and feasible.

We now know that the oldest rocks on Earth date to 4.3 billion years ago, about the oldest they could be when we bear in mind that the surface of the Earth started out as molten rock and had to cool sufficiently for rocks to form the Earth's crust, for an atmosphere of some sort to build up and for water to flow. The oldest single-celled fossils have been reported from rocks about 3.47 billion years old and the oldest possible stromatolites from rocks 3.48 billion years old, both from Australia. The exact ages and identities of these oldest fossils are under constant scrutiny, and they are heavily debated by a very active and sceptical 'origin of life' research community.

Theory had suggested that the early Earth atmosphere lacked oxygen. Indeed, the evidence from the rocks confirms this, and it's clear now that there were two episodes during the Precambrian when oxygen levels increased. Today, animals depend on oxygen to survive, and photosynthesizing plants and cyanobacteria pump out oxygen, so it seems impossible to imagine a world without oxygen. However, even today, many bacteria and other simple organisms live without oxygen; indeed, many such anaerobes die in the presence of oxygen. They live in the deep oceans or buried in the sediment, or function as fermenters in nature and in vats of beer or yogurt.

The Great Oxygenation Event happened between 2.5 and 2.3 billion years ago, when the level of oxygen in the atmosphere became detectable. Then, between 800 and 600 million years ago, just before the Ediacaran organisms diversified, oxygen levels rose to something like present-day levels. The oxygen came from photosynthesis, a fact that forms the bedrock of English scientist James Lovelock's famous Gaia Hypothesis: that life and the Earth have evolved in intimate connection ever since the origin of life. This apparently happy coexistence of Earth and life came under greatest threat when the Earth entirely froze, from the Poles to the Equator.

SNOWBALL EARTH

Hints of abundant ice through the Neoproterozoic, that last block of Precambrian time, had been noticed by geologists for many years. The crunch came in 1964 when Brian Harland, a geologist from the University of Cambridge, published a geographic reconstruction of the Earth in the Neoproterozoic showing that glacial rocks from Greenland and the Norwegian archipelago of Svalbard had formed in the tropics. If the tropics were frozen, he argued, the whole Earth must have been frozen. Joe Kirschvink, a geologist and geophysicist at CalTech in the United States, christened this 'Snowball Earth' in 1992, and the name stuck.

The evidence for glaciation includes glacial striations (scratch marks on rocks made by glaciers moving past), dropstones (rocks that fall from the bottoms of icebergs into sediments of another type), varves (layering of sediments in glacial lakes) and diamictites (ground-up rock debris made by moving glaciers, often called 'glacial till'). It's not known what kick-started the global glaciation, but there were several phases of extreme cold during a grim time between 717 and 635 million years ago.

But was the Earth totally covered with ice? Critics say that absolute ice cover would have killed off all of life, and we know that many groups of mainly microscopic living organisms, in fact, survived. Could 'Snowball Earth' actually have been 'Slushball Earth', where the equatorial strip never completely froze over and some life survived

Imagining Snowball Earth. A modern world is progressively covered with ice, from poles to Equator, just as might have happened in the Precambrian.

in that belt? A number of rock successions that span this period appear to show switches between times of heavy glaciation and inter-glacial episodes, when climates warmed and the ice retreated.

Whether snowball or slushball, it has been suggested that this caused the first mass extinction on Earth. We can't quantify its impact because we hardly know what life was like beforehand. We do know for sure, however, that life of all kinds blossomed after the end of the last of the Neoproterozoic ice ages.

... BUT WHAT KINDS OF ANIMALS WERE THE EDIACARANS?

Since 1947, palaeontologists have assigned the Ediacaran beasts to nearly every imaginable animal group, or even their own kingdom. On the more outlandish end of the spectrum, some researchers have seen them as more fungus than animal, or even as a tribe of quilted creatures, like some sort of antediluvian airbed, unlike anything now living.

Sprigg argued that his Ediacaran fossils were jellyfish. The first fossils of the frond-like *Charnia* were identified as sea pens, a kind of coelenterate related to corals. Some modern sea pens are fixed to the seabed and grow as branching frond shaped structures, just like *Charnia*, but these modern creatures tend to live in deep waters, and their branches are separate and tentacle-like, whereas *Charnia* appears to be a more solid structure without gaps between the branches. The segmented *Dickinsonia* has been identified as a jellyfish, a coral and a worm, among other things. *Tribrachidium* was seen by some as an echinoderm, something like a sea urchin. *Spriggina* has been called a worm or an arthropod. *Kimberella* has been identified as a mollusc, even an early slug-like animal.

Two other hypotheses are that the Ediacaran fossils are not related to modern animals at all, but belong to their own unique group or groups. In 1989, the German palaeontologist Adolf Seilacher (1925–2014) named a new phylum for these beasts: Vendozoa. He argued that they all shared a unique pneumatic structure where the outer skin was thick and flexible, and the interior was filled with fluid. Many of the organisms were quilted, having compartments and internal tie points to maintain their shape, and they were strong, flexible and capable of changing shape as they moved or were moved around. What the vendozoans actually were, though, puzzled Seilacher: overblown microbes, fungi or even lichens?

Most researchers agree that it is hard to allocate the Ediacaran fossils to modern groups, but one thing is for sure: their basic anatomy – some circular, some bilaterally symmetrical (where the left and right sides are a mirror image of each other) – is what would be expected in an assemblage of ancestors of modern animals of this age. How do we know that?

THE ORIGINS OF ANIMALS

The answer comes from the DNA of modern animals. Animals range from sponges to sea urchins, corals to crabs, sea pens to humans. Over the centuries, naturalists have struggled to understand the relationships between the major animal groups, or phyla. A phylum is a major group like Mollusca, such as snails or squid, or Arthropoda, such as crabs and insects. Generally, biologists can assign any animal to its phylum because each phylum has distinctive characteristics. Mollusca, for example, generally have a shell made from calcium carbonate and a soft body that can curl up inside the shell. Arthropods have multiple jointed limbs and an outer skeleton made from chitin. But are arthropods more closely related to worms or to molluscs? Are vertebrates (backboned animals like us) more closely related to sea urchins or to worms? There were some clues to be found in anatomy; for example, vertebrates and sea urchins share specialized features of their early embryos and patterns of development. But the anatomical evidence soon ran out and biologists struggled to find fundamental shared features that would indicate close relationships between many of the major animal groups. The evolutionary tree was drawn as a series of lines connecting back to a very large question mark deep in the mists of time.

DNA sequencing changed all that. With the advent of molecular techniques to compare the DNA sequences of organisms in the 1960s came the realization that the amount of similarity, and particularly the details of that similarity (for example, shared segments of the DNA or genes), can help identify actual relationships between species in a reliable manner. For example, early analysts noted that the DNA of humans and chimpanzees was nearly identical, whereas the DNA of a bumblebee and a human was very different. It soon turned out that the classical anatomists had been right to pair vertebrates and echinoderms, but some other groupings were unexpected.

One such grouping was the Ecdysozoa. Biologists had long known that many animals shed their external skeletons as they grow (ecdysis), especially arthropods such as insects, spiders and crustaceans, which shed ghosts of themselves many times during their lives. Other animal groups that shed their external skins in this way include the nematode worms, onychophorans (velvet worms) and priapulids (penis worms). This was an anatomical clue that had been ignored by early biologists because they thought that skin shedding was something many animals could do; in fact, it was a definite innovation by the ancestors of Ecdysozoa.

Identifying the shape of the evolutionary tree is one thing; dating it is another. In the early days of DNA sequencing in the 1960s, analysts assumed there had been a strict molecular clock, meaning that molecular evolution went at a constant pace. If this were true, then the tree could be dated by the amount of molecular difference between species, with 1% difference, for example, counting for 10 or 20 million years. However, it is now known that all rates of evolution are variable, and molecular evolutionary trees are calibrated (that is, dated) using fossils of known position in the tree and known geological age. This is why it is very important to determine, for example, whether *Kimberella* is a true mollusc, a member of the group comprising all the modern forms, or located somewhere deeper on the tree before the point at which the modern groups diverged. Darwin would have loved the present position of research on the early evolution of animals and the process of making and dating the tree, which involved using all the diverse tools at our disposal, including anatomy, gene sequencing and fossils.

The evolutionary tree was reshaped time and again by new molecular studies. We start with sponges and corals, at the bottom of the tree, and then there's a big event: the origin of animals with bilateral symmetry, called formally the Bilateria. These are all the worms, arthropods, molluscs, vertebrates and so on. Even though a starfish has 5-rayed symmetry, it is still bilaterally symmetrical as well. These new genomic trees proved that the deep evolutionary divergences between fundamental groups of animals, such as sponges, corals, bilaterians and ecdysozoans, happened in the Precambrian, from 660 to 540 million years ago, just at the time of the Ediacaran beasts. The tree is dated by some very early confirmed fossils of sponges and corals, and the other dates have to fall into place, with some wiggle room, according to the geometry of the evolutionary tree based on DNA of modern animals. So, even without fossils, we know all the modern animal groups originated before the dawn of the Cambrian. But where did the Ediacarans go?

THE PRECAMBRIAN–CAMBRIAN MASS EXTINCTION

Perhaps some of the Ediacaran beasts went nowhere and truly are the ancestral forms of corals, jellyfish and worms. To determine whether there was a mass extinction, we have to be sure that these extraordinary animals did in fact die out. A counter-suggestion is that a window of exceptional fossil preservation closed and that although the animals lived on, their fossils are not found. This could relate to profound chemical and physical changes. First, the increase in oxygen levels near the end of the Proterozoic would have made some of the chemical modes of preservation much harder, especially the replacement of soft tissues by iron pyrites (see page 106). Second, new groups of animals adapted to burrowing and so might have physically perturbed carcasses on the seabed and broken them up.

It seems more likely, however, that the Ediacaran animals really did disappear. There is no evidence for a sudden crisis – such as an asteroid impact or a huge volcanic eruption. Perhaps the large, soft-bodied Charnia and Dickinsonia fell prey to new groups of predators – animals with hard shells and sharp jaws that could snap them up. Indeed, the Cambrian began with an explosion of such mobile, armoured creatures.

The Precambrian–Cambrian mass extinction was one of the most significant that Earth has seen, separating the pre-extinction Ediacaran animals, with all their weird and wonderful properties, from the definitely modern animals of the Cambrian. Although it is not known what happened during the earlier Snowball Earth times, our knowledge of the history of life, in all its twists and turns, improves massively in the Cambrian, a time that was topped and tailed by extinction events, but also marked by hugely creative diversification events.

CHAPTER 2

The Cambrian Explosion and Extinctions

HAIKOU AND THE WORLD'S OLDEST FISH

In April 2013, I was part of a team of Chinese geologists visiting Yunnan Province in southwest China, near the huge Yuxian Lake, looking for specimens of the world's oldest fish. We were in search of examples of the Chengjiang Biota, an assemblage of beautiful, exceptionally preserved fossils from the Early Cambrian, which had been found in 1984 and offered amazing new insights into the early evolution of animals.

The Haikou quarry on the lake's east shore, which overlooks a narrowing of water (Haikou means 'lake mouth'), is full of rock shards from earlier quarrying operations that are pale yellow in colour; any fossils stand out clearly in brown, black, red and blue. We set to work in the quarry, turning over thin, hand-sized slabs of slaty rock and splitting them here and there, looking for traces of life. I found only a few odds and ends, but Professor Shixue Hu, from the Chengdu Geology Center and leader of the trip, found dozens of fossil specimens – this was his old stomping ground when he had worked for the Yunnan branch of the China Geological Survey. Although we didn't find any fossil fishes that day, such specimens have been found at this spot, and one of them is named *Haikouichthys* ('Haikou fish') in honour of the site.

Haikou is one of several localities where fossils of the life of the Early Cambrian, dated as 518 million years ago, have been discovered, and these are very different from those of the Ediacaran. By now, the 'Cambrian explosion' had happened – a time when enormous numbers of new groups of animals emerged, including our own earliest ancestors. It was also a period of experimentation when evolution generated some astonishing animals that are hard to comprehend and certainly hard to match with any living forms. By the end of the Cambrian, however, many of these evolutionary experiments had disappeared.

It is necessary to understand not only the Cambrian explosion but also the Late Cambrian mass extinction event and its role in weeding out the groups that survived in order to understand modern life. American palaeontologist Stephen Gould (1941–2002) suggested that the life of the Cambrian was so otherworldly and rich that it somehow exceeded the range of forms that are seen subsequently and today. Is this true? And just how important are the famous locations, such as Chengjiang in Yunnan Province and the Burgess Shale in British Columbia, Canada, where fossils are exceptionally well preserved? We will explore these themes to discover why the Cambrian is such an intensely interesting time in the history of life.

THE CAMBRIAN EXPLOSION

The American geologist Preston Cloud referred to the 'Cambrian radiation' in 1948, and Adolf Seilacher, the key proponent of the Vendozoa hypothesis for Ediacaran organisms (see page 29), spoke of an 'explosion' of the traces of life at the beginning of the Cambrian. By the 1970s, these two expressions had been combined as the 'Cambrian explosion', meaning the rapid diversification, or radiation, of new forms of animals about 540 million years ago. As we saw in Chapter 1, Charles Darwin had noted the apparently sudden appearance of rare trilobites and other fossils at this time, and the evidence that he was right has grown enormously since.

Popular books by Gould and the University of Cambridge palae-ontologist Simon Conway Morris, as well as the work of Morris's supervisor, English palaeontologist Harry Whittington, in the 1960s and 1970s, have led to wide interest in the Cambrian explosion, fuelled by the earlier discovery of the extraordinary fossils of the Burgess Shale in Canada and, from the 1980s, the Chengjiang fossils in China. These exceptionally well-preserved fossils from China and Canada, as well as Greenland and other localities, are yielding up ever more detail about the origins of major animal groups.

Some researchers have argued, however, that the Cambrian explosion did not happen and is merely an artefact of a very patchy fossil record. This is something like Darwin's observation that fossils were apparently absent in the Precambrian because we hadn't looked hard enough, or they were otherwise hidden in some way, hinting that with further work such ancient fossils ought to be found. Maybe the conditions of fossilization had somehow changed and so fossils began to be found in the Cambrian, even though the late Precambrian had also been a time of teeming and rich life. However, as palae-ontologists hunt for fossils worldwide across the Precambrian–Cambrian boundary, they find that the sequence of appearance of major fossil groups is always the same. The fact that there is a pre-dictable, stepwise appearance of different beasts looks more real than artificial and indicates that the Cambrian explosion did occur.

According to Graham Budd, professor of palaeobiology at Uppsala University in Sweden, this period of expansion and diver-sification can be divided into four stages. First came 'Tube world', 550–536 million years ago (Ma), which spanned the Precambrian–Cambrian boundary; this was when strange, tiny fossils such as *Cloudina* existed in abundance. Made of five to ten upended cones, fitting one inside another, *Cloudina* resembles a very rough stack of plant pots and presumably wobbled in the seabed currents as it drew organic particles into its tube-like centre. At this time, other animals were beginning to burrow downwards into the seabed sediment, not just move around on the surface.

Next came 'Sclerite world' (536–525 Ma), which is characterized by small shelly fossils: tiny shells or scale-like structures called sclerites that have been identified all around the world for this time period. For years, palaeontologists have struggled to identify exactly what these little cones, tubes and grooved plates, mostly 1–2 mm across, might be – perhaps shells of individual tiny creatures or parts of some sort of chain-mail-like armour on worm-like creatures.

The third stage of the Cambrian explosion, according to Budd, is 'Brachiopod world' (525–521 Ma), which was dominated, not surprisingly, by brachiopods (seabed filter-feeders with two-valved shells). One of the great animal phyla that originated during the Cambrian explosion, brachiopods still survive today, although they are now quite rare. Sometimes called 'lamp shells', because they look like small Roman oil lamps, they have one larger valve with a small hole at the tip, overlain by a smaller, lid-like shell.

Budd's fourth stage is 'Trilobite world' (521–514 Ma), when trilobites diversified. A famous extinct group of arthropods, trilobites are three-lobed creatures whichever way you look at them – three series of segmented skeletal plates longitudinally, and a head–body–tail tripartite division from front to back (see Plate V). Ranging in size from smaller than your little fingernail up to a gigantic 45 cm (18 in.) long, there were 20,000 species and they were key predators in the oceans for 290 million years from the Cambrian to the Permian.

THE BURGESS SHALE

To see the beginning of scientific and public excitement about the Cambrian as a time of revolution, we have to go back to 1909. It was then that Charles Walcott of the United States Geological

The famous Burgess Shale fossil site in the Rocky Mountains of British Columbia, Canada; fossils are found in the shaly slabs.

Survey, whom we met in Chapter 1 swerving down the Colorado River on a raft in 1883 and reporting the first Precambrian stromatolites, made a great discovery while engaged in fieldwork in the northern Rocky Mountains in British Columbia, Canada. When examining black shales of Cambrian age high on a mountainside, he chanced on some surprising fossils: silvery-grey traces of the limbs, bodies, heads and eyes of arthropods, which showed their soft parts as well as their skeletons. Sensing he had found something extraordinary, Walcott returned in 1910 with his family, and they set to with picks and hammers. At the end of the season, the Walcott family had retrieved thousands of samples. Walcott kept returning until 1924; by then, he and his family had extracted 65,000 specimens, which are now stored in the collections of the Smithsonian Institution in Washington, D.C.

During his lifetime, Walcott published dozens of papers describing the wonderful new fossils of the Burgess Shale – some of them trilobites, worms and molluscs that were known from other locations, but most of them representing astonishing, otherworldly creatures that were hard to classify – although he struggled to do all the research needed and to convince people of their importance. Then, in the 1960s, Harry Whittington (1916–2010) reopened the excavation sites and began a process of meticulous description, devoting more time to the work than Walcott could and also using new photographic methods to extract maximum information. His publications, and those of his students Simon Conway Morris and Derek Briggs among others, showed people the astounding anatomical information locked in the fossils.

Gould was excited about the work, and his 1989 book *Wonderful Life* made Walcott, Whittington and the others into (perhaps reluctant) heroes and brought the Burgess Shale to the millions. Gould began research as a brilliant student at Columbia University in New York, then spent most of his professional career at Harvard, before dying far too soon at the age of sixty. He wrote many papers, including three hundred essays in *Natural History* magazine, and published twenty books, all of them bestsellers, popularizing palaeontology among a general audience (although many palaeontologists hated him for that). He argued against sociobiology, genetic determinism, gradual evolution and many once-valued ideas. People paid attention when he spoke and he frequently deployed the fossil evidence, and especially appropriate numerical methods, to make his point.

The Burgess Shale dates from the Middle Cambrian, about 508 million years ago, and the cast of characters astonished the world (see Plates IV, VI). Most amazing was the huge *Anomalocaris*, perhaps 2 m (6½ ft) long, an arthropod with segmented armour over its long body, a branching tail fan, and broad leaves along the side of the body that possibly rippled in life and paddled the animal along. When the fossils were redescribed by Whittington and Briggs in 1985, they showed that *Anomalocaris*, which means 'weird shrimp', consisted of a number of previously misidentified portions. Attached to the front end were two long, curved articulated arms that could curl downwards, with arched chitin scutes (plates) that fitted together like the leg and knee shields of a suit of medieval armour. These great arms had been misidentified earlier as individual curved worm-like creatures, but Whittington and Briggs put them back in their proper place. Another misidentified fossil resembling a pineapple ring – a circular structure with a hole in the middle and numerous radiating segments – had been called a jellyfish, but in fact was the mouth of *Anomalocaris*, located beneath the head and in reach of those great curved arms. The anomalocarids were predators, feeding on prey 2–5 cm (1–2 in.) across – most of the other Burgess Shale organisms – which they grabbed with their two feeding appendages, curled downwards to hug the prey tightly and perhaps even crush, then stuffed into the mouth. This might have functioned to further mash the prey by grinding it with the radial mouthparts.

Among the humbler seabed creatures of the Burgess Shale, *Wiwaxia* looked like a 5-cm (2-in.)-long slug covered with loosely fitting, oval-shaped plates over its back, and with two rows of seven vertical flattened spines along its back. Its mouthparts were underneath and it probably fed by grazing on organic matter; indeed, it appears to belong to an unusual group of molluscs. Like *Anomalocaris*, *Wiwaxia* was first identified from isolated parts, often flattened in the rock, and only over the years, and with much hard work, have the parts been put together to visualize the complete animal.

Even more preposterous was *Hallucigenia*, the name somehow reflecting its dream-like appearance. In the first detailed studies, it was reconstructed upside down, with a row of fleshy limbs pointing upwards and its dorsal spines downwards. Once the seven pairs of wobbly-looking legs and seven pairs of stiff spines were recognized as such, it was flipped through 180 degrees and viewed correctly. *Hallucigenia* was 0.5–5.5 cm (¼–2¼ in.) long,

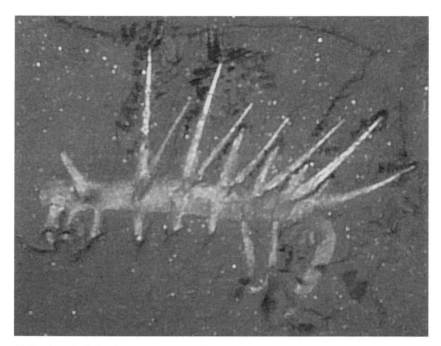

Hallucigenia. Paired spines form V-shaped structures down its back, perhaps for defence, and its short fleshy limbs are below.

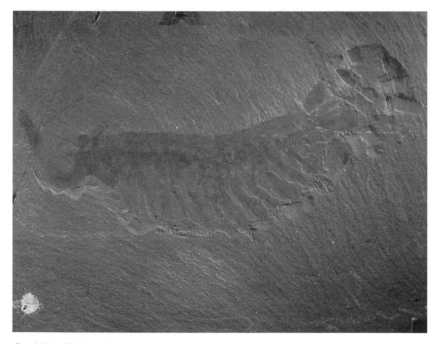

Opabinia. This arthropod has broad chitin flaps down its sides, perhaps for swimming, and five eyes and a feeding snorkel on its head (left).

with a tubular body and a long head with two eyes at the front. This vaguely worm-like beast is most probably a lobopod, a modern group that includes velvet worms, which are related to arthropods.

Another creature, with a body similar to that of *Anomalocaris*, was the 10-cm (4-in.)-long arthropod *Opabinia*. It was a three-lobed animal, consisting of a central elongate body protected by segmented armour rings from front to back, and lateral lobes for swimming. Its head end was strange: it had five eyes on stalks, arranged symmetrically on top of the head, a long, flexible proboscis tube below, shaped like the hose of a vacuum cleaner, and a pair of pincers, each bearing five spines on each side. It probably moved the muscular proboscis around in the mud probing for prey animals and when it encountered something, gripped the prey with its pincers, piercing the body with the spines, and then crushed and slurped up the delicious morsel.

Nearly two hundred species of animals have been named from the Burgess Shale – what are we to make of such a diversity of extraordinary creatures? They had never been reported from any other site, so without the Burgess Shale rocks, nothing would have been known about them. Were these weird animals unique to the Burgess Shale locality, a restricted and astonishing outburst of creativity to be seen only in that small area of ocean floor, now manifested in the Rocky Mountains of Canada? Or did these amazing beasts live everywhere in the Cambrian and were just not preserved elsewhere?

THE CHENGJIANG BIOTA

In 1984, the discovery of the Chengjiang fossil assemblage, or biota, in Yunnan Province answered these questions. In fact, although rather older at 518 million years ago, and falling in the latter stages of the Cambrian explosion, Chengjiang did share some of the same wonders found in the Burgess Shale, such as anomalocarids with their hunting arms, some other arthropods and *Hallucigenia*.

So far, over two hundred species have been identified in the Chengjiang fossil beds, and some of the most important include the oldest fishes, such as *Haikouichthys*, which we were seeking to catch when we visited in 2013. This little beast was named in 1999 and since then dozens of specimens have been found. A tiny, flat, penknife blade of a fish, only 2.5 cm (1 in.) long and with paired dark eyes at its front end, its body was pointed at front and back. All along its back and curling down beneath the tail it had a midline expanded fin, a thin membrane supported on some tough supporting rays. Inside its body, W-shaped muscle blocks powered its swimming by the sideways bending of body and fin. Behind the eyes and in the throat region were up to nine gill slits.

All these features spell out fish, or more properly vertebrate. The most basic, simplest design for a fish is like this, and *Haikouichthys* was clearly adapted for nippy swimming like a modern stickleback. The regular muscle blocks, the tail, the dorsal fin and the gill slits in the throat region are all features of vertebrates and, indeed, of their wider inclusive group called the chordates. It is immensely exciting and important to realize that vertebrates, the group to which we belong, have their origins in the Cambrian explosion. We may think humans are pretty nifty creatures compared to the other animals, but our phylum originated at the same time as the worms, molluscs and arthropods.

THE SIGNIFICANCE OF THE SKELETON

The Cambrian explosion was arguably the biggest reorganization or revolution in animal evolution: before the explosion, the strange rooted and slowly flopping and bulldozing creatures of Ediacara; afterwards, molluscs, arthropods, the first fishes, colony-building creatures and others that made burrows in search of food or for protection. Soon, as shown at Chengjiang and the Burgess Shale, there were giant predatory arthropods as large as dogs feeding on everything in sight.

Before identifying a possible reason for this expansion of new life forms, we need to consider that first, the Cambrian explosion was not a single or rapid event, but in fact lasted for 36 million years from 550 to 514 million years ago and consisted of the four stages described by Graham Budd. Therefore, there might not be a single cause for this explosion of life, such as a sudden change in ocean chemistry or temperature; each stage may have been triggered by a similar event or might have had its own distinct cause. Second, the world was also changing hugely at the end of the Precambrian and the beginning of the Cambrian, and it is difficult to pin down exact timings. For example, the Earth was undoubtedly more suitable for life than it had been, with the ending of the Snowball Earth conditions and the rise in oxygen levels during the Neoproterozoic, but these events happened about 717 and 635 million years ago, long before the beginning of the Cambrian explosion.

All of the above notwithstanding, it has often been argued that the origin of skeletons at this time must hold a clue. During the four stages of the Cambrian explosion, as many as fifty new animal groups appeared in the oceans that all had skeletons of some kind. The term 'skeleton' doesn't only refer to the internal bony skeleton of humans and vertebrates in general, although that was one of the great innovations of the Cambrian; it can also be external and found in invertebrates. The vertebrate skeleton is made from calcium phosphate, or apatite; ten or twelve other animal groups independently acquired skeletons rich in phosphate, including some brachiopods and wormy creatures, as well as many of the small shelly fossils mentioned earlier. In these, the phosphatic skeleton is external – either a shell that entirely covers the body, or a mix of smaller protective shields or internal skeletal portions such as jaws. Three other skeleton types emerged during the Cambrian explosion, two of them comprising different chemical variants of calcium carbonate called aragonite and calcite. These types of skeletons are seen in molluscs, worms, some brachiopods, trilobites and echinoderms. The third skeleton type was constructed from silica, the fundamental material of sand and glass, and seen in sponges.

Why have a skeleton at all? External skeletons are generally for protection, enabling their possessors to shrink back inside or batten down the hatches when danger looms. They can also provide support and points to which muscles can attach, as in vertebrates and arthropods, for example – the flesh of a crab or lobster comprises mainly muscle, and these muscles attach to the external skeleton to allow the limbs to operate, just as our muscles are anchored to the different elements of our internal skeletons. Skeletons also enable colonies to build up, such as coral reefs, where multiple organisms living side by side, each in their own calcite skeleton house, may enhance the survival chances of the whole colony by sharing functions.

Skeletons can be a source of minerals for the functioning of animals: vertebrates retrieve calcium and phosphorus from their bones when needed – for example, when female egg-layers build egg shells, or when heavy exertion is needed (phosphorus is part of the physiological energy cycle). Mineralized skeletons indicate a need for minerals, and geochemists have sought evidence that the world's oceans in the Early Cambrian became enriched in suitable chemical ions for the different fleshy and naked ancestral animals to absorb and use in building skeletons.

At one time, geochemists argued that there had been a 'Strangelove ocean' at the beginning of the Cambrian, characterized by low levels of oxygen and absence of life; it was named after Stanley Kubrick's 1964 movie Dr Strangelove in which the lead character was intent on destroying all of life. Such an ocean would have led to great loss of life, especially among the green microorganisms that constructed stromatolites (see pages 23–24), so eventually new forms of life would have exploded to fill the emptied ecological space. However, no evidence for such an event has been found.

There is, however, evidence for raised levels of calcium and phosphorus, both elements eroded from rocks and distributed in ocean waters as a result of high levels of perturbation and overturn (when shallow and deep waters replace each other completely). Oxygen and carbon levels in the oceans varied substantially during the Early Cambrian, and some of the sudden rises and falls appear to match bursts of evolution, but whether the chemical changes in the oceans drove the evolution of new animal groups or the burgeoning of new animal groups affected the cycling of carbon and oxygen, acting in the opposite direction, is hotly debated.

Most researchers see the Cambrian explosion as a drawn-out affair, extending minimally over 36 million years, but possibly much longer, as part of a process that encompasses the Ediacaran organisms found at the end of the Precambrian and extends through Budd's 'Tube world' and the rest, so it would not have been necessarily caused by a single chemical trigger that occurred at one particular point. Whatever the cause or causes, however, the consequences were astonishing in terms of the diversity of new forms of animal life.

TRIMMING STEVE GOULD'S HEDGE

Delving into the Cambrian gives a thrilling insight into big events in the history of life. When fossils are exceptionally well preserved, as in the Burgess Shale and at Chengjiang, amazing creatures have been discovered and we can see another world far from our own. In the same way that determining exactly what caused the Cambrian explosion is difficult, so is interpreting the runaway evolution of the Cambrian. Was it possible that in some way the Cambrian explosion was a special event, a time when evolution was running in overdrive, or was it perhaps something a little more constrained?

Steve Gould certainly believed the former. In his book about the Cambrian explosion, *Wonderful Life*, he argued that conditions in the Cambrian were so ideal for life and there was so little diversity of life forms beforehand, that it was a time of experimentation. Animal groups multiplied, some of them familiar to us today, but many of them strange and unexpected, as we have seen. Based on this, Gould proposed a new model for evolutionary diversifications, times when major new groups of organisms burst on the scene. He likened the evolutionary tree of the Cambrian to a tangled hedge, branching and splitting wildly, sending out shoots and branches in all directions. Each branch represents a new species. He argued that the Burgess Shale fauna included more extremes of body shapes and functions than have been seen since.

Gould envisaged this unruly Cambrian growth necessarily having to be brought under control, and suggested that what happened at the end of the period was a partial clipping of the hedge. As the experimental groups diversified, they would have come into competition, and some of them died out. Only the well-adapted, resilient creatures could survive the extinction event in the Late Cambrian. The great evolutionary hedge trimmer clipped off many of the branches and only those lines that led to the modern animal phyla survived.

This view has been criticized as perhaps a little overwrought. Although the Cambrian was a time of experimentation, when palaeontologists assessed just how peculiar the Burgess Shale arthropods were, they found they were no more bizarre than living arthropods; in fact, the range of body forms among modern crabs, lobsters, centipedes and insects is huge. Single locations in the Brazilian rainforest doubtless show as much diversity of shape among beetles, butterflies, spiders and weevils as in the Burgess Shale. It is true, however, that although many of the Burgess Shale animals survived into the Ordovician, many of them did in fact disappear in the Late Cambrian.

THE SPICE EXTINCTION

It is hard to pin down the exact timing and impact of the Late Cambrian extinctions. Although more is known about these events than the impacts of Snowball Earth and the late Precambrian mass extinctions, they remain a little mysterious. Previously, palaeontologists believed there were five or six sharp phases (pulses) of killing, but attention has now focused on a single cycle of death, perhaps lasting for three million years and associated with geochemical anomalies known as the Steptoean Positive Carbon Isotope Excursion (SPICE).

It has long been known that there was a sudden break in the history of life across the boundary between the Cambrian and the overlying Ordovician, dated at about 485 million years ago. In the 1800s, in the early days of geology, the stratigraphic boundary in the rock layers that marked this break was indicated by a major switch in the kinds of trilobites, brachiopods and other fossils that could be found. However, the main events happened ten million years earlier, as we now know thanks to improved dating of the rocks.

The Cambrian extinctions started 497 million years ago, when there was a reef revolution. Until then, reefs had been formed in shallow seas largely by microbes making modestly sized stromatolites (see page 23). These reefs were wiped out by changing ocean chemistry and falls in sea levels. New reef types formed from sponges, and combinations of microbes and sponges emerged as sea levels rose again. Indeed, this reef replacement was part of a broader series of extinctions through the Late Cambrian, some associated with the SPICE event from 497 to 494 million years ago. SPICE was a time of not only high oxygen in the atmosphere and climate cooling, but also high rates of carbon and sulphur burial and loss of oxygen (anoxia) from seabed sediments. As there was little or no life on land, the major impact of SPICE was on marine life – anoxia and high sulphur levels are often associated with the death of marine organisms (see pages 163–68). Reefs were massively perturbed. There were high spikes of trilobite extinctions, and brachiopods and conodonts (lamprey-like early vertebrates), two other important groups of the Middle Cambrian, also showed major losses. We don't know what happened to all the Chengjiang and Burgess Shale organisms. Some did survive into the Ordovician, but many are never seen again, and so might well have disappeared too.

These big events of the deep past are hard to understand because of the difficulties in collecting information and fitting it to the geological time scale, but we know for certain that another mass extinction had occurred, which opened the world, and the seabed in particular, to a remarkable acceleration in evolution during the Ordovician and a big change in the way animals functioned, meaning two-dimensional ecosystems changed to a new three-dimensional world. Life would no longer be limited to the flat seabed, moving only forward, back or sideways; new kinds of animals moved upwards, building structures such as reefs, and others burrowed down into the sediment (see pages 55–56). Again, a mass extinction appears to have unlocked a door and let in remarkable new opportunities.

CHAPTER 3

Ordovician Diversification and Mass Extinction

THE GREAT ORDOVICIAN BIODIVERSIFICATION EVENT

Geologists label things. Once something has a name, it becomes a thing and can be discussed. The Great Ordovician Biodiversification Event, or GOBE for short, is one such, but it had long been masked from view. Most palaeontologists saw the Cambrian explosion as a time when all the marine animal groups emerged and when the fossil record of the oceans became suddenly rich, and this seemed to last through the Cambrian and into the Ordovician. When proper statistical analyses of data were conducted, there was a significant improvement in the understanding of the timescale, much of it thanks to the work of American palaeontologist Jack Sepkoski (1948–1999), working at the University of Chicago.

In the 1980s, as a graduate student, Sepkoski set himself the unenviable task of collecting all the information available on the occurrences of fossil animals worldwide. Up to that point, there had been a few such efforts, but they were often based on hastily compiled information, and often at high taxonomic levels such as classes or orders. Sepkoski wanted to look at the origins and extinctions of families and genera, lower down the hierarchy of classification, realizing that such an investigation would be too ambitious at the species level because the numbers would be huge (each genus can contain one to fifty species, each family one to fifty genera, and so on) and that the exact nomenclature of species and their distributions in geological time is often disputed.

As he started publishing his graphs of fossil diversity in the oceans through time, Sepkoski noticed a very clear break in the extended Cambrian explosion, separating the Cambrian from the Ordovician. There was a distinct Cambrian step when marine life diversified to about a hundred families and five hundred genera, followed by a second step in the Ordovician when levels jumped to five hundred families and fifteen hundred genera. He identified a Cambrian fauna of small shelly fossils, simple brachiopods, trilobites and reef-building sponges called archaeocyathans. The earlier idea that the Cambrian explosion spanned through the Cambrian and Ordovician could be refined: this astonishing diversification event in fact consisted of two phases, one that we have already explored in Chapter 2, which gave rise to many modern groups, as well as uniquely strange creatures; and a second Ordovician pulse that relates in particular to the denizens of modern oceans.

The GOBE lasted for most of the Ordovician from 485 to 444 million years ago, and different animal groups expanded at different points during this span. First came new types of plankton, including graptolites: colonially living, floating fossils resembling small sawblades. Each zig and zag along the blade, even though only millimetres across, was home to a little creature that plastered and maintained its house at microscopic scale; the zigs and zags distinguish the species. Graptolites became important plankton for the next 235 million years, only disappearing at the end of the Permian. Then came brachiopods, which had existed

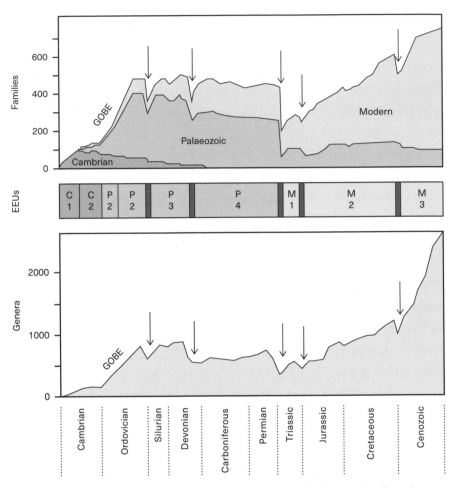

Sepkoski's famous graphs of the diversification of life in the sea: families above, including Ecological Evolutionary Units (EEUs, representing assemblages of animals that characterize particular spans of geological time), and genera below. The 'big five' mass extinctions are indicated by arrows.

through the Cambrian, but now flourished as a great many new groups, all with distinct shapes and shell ornaments. As we will see, they were major seabed organisms, indicative of geological time, but also marking out assemblages at different water depths according to the species present.

Corals diversified later in the Ordovician. There had been reefs in the Cambrian (see page 48), but these were composed of archaeo-cyathan sponges, microbes and other organisms. From the

Ordovician onwards, the world's oceans became home to reefs formed from corals. Sponges and other groups were still present, but corals now made the major reef structure. Reefs are important because they are megastructures that have a huge ecological impact, in some ways like a tropical rainforest. If a reef is wiped out, just as if a forest is destroyed, you are left with a kind of 'desert', a patch of seabed (or land) that supports very few species. As the reef constructed itself, often over hundreds of years, not only did it contain many coral species, it also provided a remarkably rich habitat for other species: trilobites and early sea urchins crawled through the structure, while other animals like brachiopods and crinoids (sea lilies) attached themselves on and around the reef. Yet others, such as cephalopod molluscs with their shells and early fishes, swam around and above the reef and in and out of its crevices, seeking out prey. Reefs, like tropical rainforests, are important because they can represent hotspots of exceptionally high biodiversity.

THE PLANKTON REVOLUTION

Plankton are the plants and animals, usually tiny, that float in surface waters of the sea. The plant-like plankton (phytoplankton) drive the ocean food system by generating their cells and tissues through photosynthesis, extracting energy from the Sun and minerals from the seawater. Phytoplankton are fed on by zooplankton, and in turn, the zooplankton by larger creatures about the size of your thumb, and they by ever larger forms. Many fish, for example, feed on these floating food sources, as do the largest marine animals of all, the baleen whales, which survive on krill (shrimps that feed on the microscopic floating life of the oceans).

At the beginning of the GOBE, there was a plankton revolution. Cambrian marine life was focused on the seabed; nutrients in the form of organic particles arrived there from dead organisms and sometimes washed in from the land. At the end of the Cambrian, the first phytoplankton emerged and diversified through the Early

Ordovician. This then provided food supplies for zooplankton groups, most notably the graptolites, as well as others called chitinozoans and radiolarians. Chitinozoans appear in the fossil record at this time, and continued as important plankton through the Ordovician, Silurian and Devonian periods. They are tiny, flask-shaped organisms, mostly of microscopic size, that have sometimes been identified as eggs or juveniles of graptolites and sometimes as a distinct group. Radiolarians exist today and are delicate single-celled creatures, often with beautiful, delicate skeletons made from silica. They were and are important and abundant zooplankton that mainly survive by feeding on phytoplankton.

The explosion of plankton as the first step in the GOBE might appear insignificant compared to the diversification of larger animals such as trilobites and molluscs because it happened at a microscopic scale. However, it in fact marked a profound change in Earth-Life history, and most notably to the carbon cycle. In the absence of life, carbon enters the sea through erosion of rocks and transport down rivers, and is then buried on the ocean floor where it can eventually be drawn under moving tectonic plates; there the rock is melted and the carbon can pass back into the atmosphere through volcanic eruptions. When plankton evolved, the process speeded up, with carbon from the atmosphere being captured by photosynthesis in plankton cells and eventually transported to the seabed in showers of their tiny dead carcasses. The plankton revolution therefore brought about the most important and long-lasting of the changes affecting the Earth during the GOBE.

WHAT CAUSED THE GOBE?

Just as palaeontologists have tried to understand the phases of the Cambrian explosion in terms of changes in the Earth and ocean chemistry, they also have been keen to identify external causes of the GOBE. The Ordovician was certainly a time of active Earth movements, with the continents and shallow seas jigging majestically, over millions of years, to new latitudes. These movements in the layout of the Earth's crust were accompanied by periodic, but intense, volcanic activity in many places. Volcanic eruptions pump greenhouse gases into the atmosphere, raising temperatures worldwide and high temperatures, providing they are not too high, can be associated with rising levels of biodiversity. Volcanoes also pump minerals into the oceans, and sometimes these minerals are useful for shell and skeleton growth (see pages 44–45). Volcanic activity and continental movement also generated new opportunities for ocean life by generating new tracts of continental shelf and creating divisions between others. Topographic divisions such as mountain chains on land and barriers across oceans mean species cannot move freely and the populations on either side of the barrier may go their own ways, evolutionarily speaking, leading to distinct species and an overall increase in biodiversity.

Oxygen levels were rising, too, and this could have enabled some groups of animals in the oceans to flourish. The warming generated by volcanic activity was also interspersed with cooling episodes, some of which might have generated some opportunities in tropical regions that would otherwise have been too hot for life. Heating and cooling can be good for life at times, but, as we saw with Snowball Earth in the Precambrian, and will see in many of the later mass extinctions, sharp or large temperature changes can lead to catastrophic loss of life.

We have already seen that life can beget life: a reef in itself marks a rise in biodiversity, but the reef then offers opportunities for biodiversity to increase ten-fold among other animal groups. The huge increase in the diversity of life during the GOBE was perhaps also enabled by another major structural change: the shift from two to three dimensions. Cambrian life in the sea was limited to the seabed, whereas Ordovician life conquered the third dimension more convincingly. New groups of animals began to use more of the water column, with some swimming in open water, some floating as plankton and some burrowing, penetrating into the sea-floor muds and sands to find food.

All those amazing animals from the Burgess Shale and Chengjiang only walked or bulldozed around within a narrow depth zone of a few centimetres. They were not exploiting all the food resources above and below. As the Ordovician began, numerous groups of worms, arthropods and others started exploring opportunities in a downwards direction. Palaeontologists know this because abundant evidence in the form of trace fossils has been found, such as burrows and trails that show what animals, even soft-bodied animals, were doing. During the Ordovician, burrowers went down as much as 1 m (3 ft) into the seabed, opening up vast new food reserves and allowing any area of the ocean floor to support many more species than before. Swimmers and plankton extended the range right to the surface of the water.

As a consequence of these changes, food chains became longer in the Ordovician. Whereas in the Cambrian there were three or four links in the chain, from microbes and small beasts at the bottom to a lower level of small predator, then to larger predators such as the prey-hugging and crushing *Anomalocaris*, in the Ordovician, many new levels of feeding evolved, including top predators that swam freely in the oceans. One such was the cephalopod *Cameroceras*, which reached 8 m (26 ft) in length. This was one of many orthoceratid cephalopods of the period: long pointed shells with an octopus-like, many-tentacled body inside, which we assume grappled with larger predators, including anomalocarids and early fishes.

DR ELLES, PROFESSOR JONES AND MR BANCROFT

Towards the end of the Ordovician a mass extinction event occurred, which was identified in the 1960s. However, clues to its existence had been found earlier, in the 1920s and 1930s, by three remarkable British geologists at the University of Cambridge. The first of these was Gertrude Elles (1872–1960), a remarkable woman and a pioneer

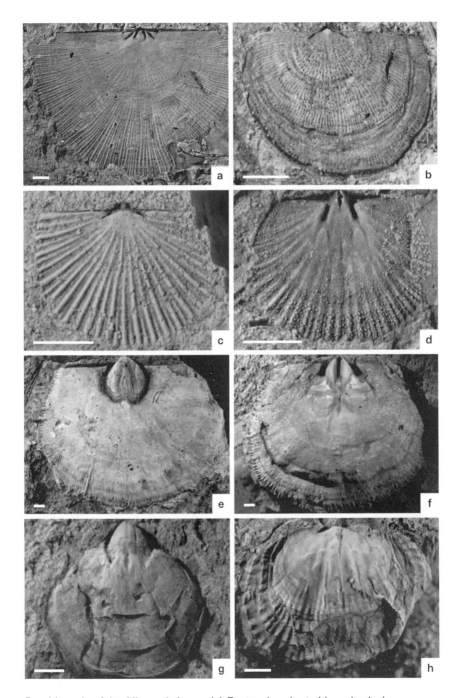

Brachiopods of the *Hirnantia* fauna: (a) *Eostropheodonta* (dorsal valve);
(b) *Paromalomena* (dorsal valve); (c) *Fardenia* (dorsal valve); (d) *Dalmanella*
(dorsal valve); (e, f) *Hirnantia* (ventral and dorsal valves); (g) *Hindella* (ventral
valve); and (h) *Cliftonia* (dorsal valve). Scale bars = 2 mm.

in acquiring a degree at a time when women did not generally go to university. Later becoming Vice Principal of Newnham College, a women's college at Cambridge University, she was famous for her large collection of hats (at that time, women at the University of Cambridge were obliged to wear them when delivering lectures). Elles recognized a unique fossil assemblage in the later-named Hirnantian rocks of North Wales, arguing that they were Early Silurian in age based on her detailed fieldwork on the rock sections of the wind- and rain-blasted hills around Cwm Hirnant.

Professor Owain T. Jones (1878–1967), then at the University of Manchester in the UK and later Cambridge, was quick to disparage her work, however, writing 'It is sufficient for the present to state that this section has been misinterpreted by Miss Elles.' He argued (correctly) that the Hirnantian rocks were in fact Late Ordovician. Jones was known for his fiery temper, and the third player, Mr John Bemis Beeston Bancroft, would soon experience his full fire. From 1930 to 1932, Bancroft was employed as a research assistant at the University of Cambridge to study and publish on Jones's Early Silurian fossil collections. However, Bancroft spent most of his time working on his beloved Ordovician fossils, the Hirnantian brachiopods. When Jones discovered this, he was furious and fired Bancroft, who returned to his family home in Blakeney, Gloucestershire, where he continued his work without an academic post and was widely shunned because of the professor's influence.

Bancroft had served as a soldier in the First World War, and through the 1920s worked on fossil brachiopods from the Ordovician of Shropshire and North Wales. He was particularly interested in straightening out problems with the relative dating of the rocks, and he realized that the abundant brachiopods might be the key. Brachiopods, which originated during the Cambrian explosion, had diversified substantially in the Cambrian and Ordovician, dominating the sea floor in many places. Some were tiny, at a few millimetres across, but others were as big as a human hand. They were usually attached to the seabed or rocks by a tough fibrous pedicle, a kind of stalk, and the two-valved shell floated in the water or, more commonly, sat firmly on the bottom because of the weight of the shell. Brachiopods fed by filtering food particles from the water using a feathery, fleshy internal sieve structure called a lophophore.

What interested Elles, Jones and Bancroft was that these early brachiopods showed a great range of shell sculpture, shell shape and size, and that they appeared to evolve in unison over wide areas, even worldwide. Therefore, a particular brachiopod species, perhaps circular in shape, with deep zig-zag grooves radiating from the attachment point of the pedicle to the edge of the shell where water entered and departed, might usefully characterize a particular half-million-year chunk of time in Wales, Sweden, Canada and Argentina.

About 1930, Bancroft had published several uncontroversial papers about his brachiopod endeavours, and then he made an important discovery. He was impressed by the work of Elles and was convinced that the unique fossil assemblage from Cwm Hirnant identified a short-lived episode of unusual environmental

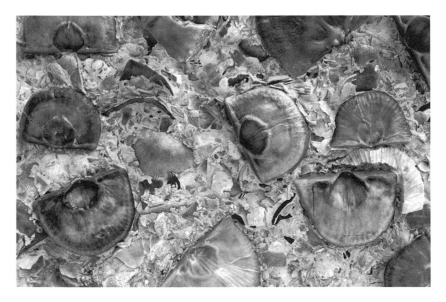

A slab of limestone containing multiple examples of *Strophenema* sp. and *Rafinesquina* sp., key brachiopods of the Ordovician Period.

conditions. Bancroft privately published his new ideas in 1933 in a short paper with the encouraging title *Correlation Tables of the Stages Costonian–Onnian in England and Wales*. At first, nobody paid any attention, but today this obscure and hard-to-find work is widely cited. Tragically, in 1944, Bancroft was killed in action in the Second World War. Despite continued opposition from Jones, the admiring young Scottish geologist Archibald Lamont (1907–1985) ensured that a lengthy monograph of Bancroft's was published posthumously in 1946 in the American *Journal of Paleontology*.

The work of Elles, Jones and Bancroft has led to a greater understanding of the importance of the Ordovician, both in terms of the explosion of life at the beginning and the remarkable extinction at the end. Although they did not have the whole picture, we now know that the Hirnantian stage can be detected worldwide – a sliver of time representing half a million years that was key to the first of the 'big five' mass extinctions, a period when the Earth froze and life died.

THE HIRNANTIA FAUNA

Although he did not know it at the time, what Bancroft had identified as the Hirnantian time interval in North Wales was also evidence for a global mass extinction. Since then, palaeontologists have found the Hirnantian fossil assemblage everywhere. Indeed, very unusually, it seems to be strikingly uniform and global in spread. Normally, such unique assemblages of species are useful age markers across a region, such as southern Europe or Australia, but elsewhere in the world there are variations and major differences.

The assemblage is known as the *Hirnantia* fauna – after *Hirnantia*, a 1-cm (½-in.)-wide brachiopod shaped like an open paper fan, with the same zig-zag crenulation of ridges extending to the rounded outer margin. Although this modest shell is not the most abundant in the *Hirnantia* fauna, it was the one named by Lamont for its Cwm Hirnant locality, so was chosen as the namesake of the assemblage and the indicator of the Hirnantian time slice. Not only is the *Hirnantia* fauna distinctive and immediately recognizable, as Bancroft predicted, it is also marked by a sharp cut-off below and above the Hirnantian level in the rocks. The earlier brachiopods and other seabed animals disappear rather sharply below the Hirnantian. Then, after a few metres of rock representing half a million years, everything changes again. *Hirnantia* and associates are gone and a whole new fossil assemblage replaces them. So, the *Hirnantia* fauna is one of the most immediately recognizable, widespread, but short-lived animal assemblages ever found. What did it mean, if anything?

Geologists thought the *Hirnantia* fauna marked a major phase of cooler seawater because it was already known that there had been a significant glaciation across the South Pole in the Late Ordovician. This explained the unique occurrence of *Hirnantia* faunas close to the evidence for glaciation and ice, but in the 1980s, the *Hirnantia* fauna was reported more globally, from North America, the Baltic areas of Europe, central Europe, Spain, Russia, China, North Africa and New Zealand. Evidently, this cold-water fauna also occurred in tropical and temperate zones. This was unexpected and suggested a global episode of freezing. The *Hirnantia* fauna, dating from 445 to 443 million years ago, also coincided with the Late Ordovician mass extinction.

EXTINCTION STEP BY STEP

This mass extinction event seems to have happened in two steps, with two high peaks of species loss, the first linked to a global phase of freezing climates represented by the *Hirnantia* fauna, and the second occurring after climatic warming and loss of oxygen on the seabed. Brachiopod species changed rapidly, and there was a reef gap (a time of limited reef development), which can be a strong indicator of perturbed living conditions in the oceans. The trilobites changed dramatically too, key predatory arthropods of the time (see Plate v).

Overall, during this Late Ordovician mass extinction, 20% of marine families died out, representing 40% of genera and 80% of species. This escalation of percentage losses reflects the inclusive nature of the classification scheme: each genus contains many species and each family many genera. So, for a family to completely disappear, all its genera and all its species have to die – a family can survive with only one species. This line of reasoning is followed for all of the mass extinctions (see pages 100–101). Although it is assumed that the remaining 80% of families at this time were substantially depleted, the Late Ordovician mass extinction, during which four in five species disappeared, was associated with a loss of only one in five families.

Brachiopods, corals, trilobites, crinoids, molluscs and many other groups were hit hard during both phases of the mass extinction. It seems that the animals that lived fixed to the seabed were more severely hit than those that could swim, although the predatory cephalopods such as *Cameroceras* also suffered huge losses as their food sources disappeared. Animals restricted to particular geographic regions also suffered extinction more readily than those that had worldwide distributions.

Graptolites show a major loss: all the typical Ordovician kinds that existed as multiple-branched forms (groups of two, four or more graptolite blades together) died out and were replaced by single-bladed types, the so-called monograptid graptolites, in the following Silurian Period. This sharp change in graptolite form was accepted by earlier geologists as a handy tool for identifying the ages of rocks (multiple-branched graptolites = Ordovician; monograptids = Silurian), but the replacement is also an important indication of the depth and seriousness of the Late Ordovician mass extinction.

The extinctions show ecological signatures. Animals living in deeper waters near the edge of the continental shelf died out first, then extinctions occurred in shallower waters, especially the reefs, which largely disappeared. Although enough species survived in restricted geographic areas for reefs to recover in the Silurian, at the height of the crisis, they had largely been wiped out.

Worldwide distributions of marine animals also showed catastrophic changes during the extinction event. Whereas there had been ten regional-scale geographic provinces of animal life before the Hirnantian, this decreased to nine, and then five, after the second extinction phase. These geographic provinces had their own regional lists of species, and so contributed in important ways to the sum total of global biodiversity. As the extinction crisis took hold, species were wiped out in huge numbers in some provinces, but in other cases, multiple provinces merged as life came under pressure.

ICE, FALLING SEA LEVELS AND WEATHERING SHOCK

Geologists had long been aware that the Late Ordovician was a time when sea levels fell substantially, perhaps by as much as 150 m (500 ft). The obvious cause was that during the Ordovician, the great southern supercontinent Gondwana, an amalgam of modern South America, Africa, Antarctica and Australia, had drifted over the South Pole. If there is open ocean at the Poles, the polar ice melts in summer and climates are acceptable for life even though they are icebound in winter during the months of darkness and cold; when there is land at the Poles, as there was in the Ordovician (and is at the South Pole today), the ice can remain unmelted all summer because of the physics of reflection and convection. Ice on land or sea has a self-preserving property, called the albedo effect: the white surface reflects sunlight and so slows down the melting. However, ice on the ocean sits above water – warmth seeps upwards by convection and the ice melts. On land, the rock beneath the ice is also frozen to some depth and this can prevent melting, even under a warm sun. Freezing oceans lead to a massive fall in sea levels because the polar ice is drawn largely from water in the oceans, so as the ice cap expands, sea levels fall – and this is what happened in the Late Ordovician. We are seeing the opposite effect today, when rising temperatures driven by excess carbon dioxide in the atmosphere are causing the ice caps to melt and we fear for the safety of people and eco-systems around the coasts as sea levels rise.

Temperatures also fell globally during the short-lived, half-million-year Late Ordovician crisis. A temperature fall of 5–10°C (9–18°F) is estimated from rock sections worldwide, although the drop was probably larger in the icebound regions. To establish past climates, geologists use an oxygen palaeothermometer to estimate ancient oceanic temperatures, comparing proportions of the two fundamental forms, or isotopes, of oxygen called oxygen-16 and oxygen-18. The numbers 16 and 18 indicate the total numbers of protons and neutrons in the nucleus of each atom. In nature, oxygen-16 is much more abundant than the heavier version, oxygen-18. When ocean water evaporates, more oxygen-18 passes into the atmosphere at cooler latitudes, so the $^{18}O:^{16}O$ ratio of the seawater falls – for example, cold locations today show 5% lower proportions of oxygen-18 than the average for ocean water. Geologists use this knowledge of the relationships of oxygen isotopes to air temperature today to scale this oxygen ratio and calculate ancient temperatures.

Low temperatures can disturb life, but the massive fall in sea level must also have had an effect. As seas withdrew miles offshore, exposing the continental shelf, areas that had been full of marine life, including reefs, were exposed to the air. It is likely that all marine life moved progressively with the water, but as sea levels fell, those shallow-water offshore areas became increasingly smaller and there wasn't enough space. Species died out. The extinction occurred in two pulses, both associated with advances of glaciation, probably with a warmer interval between.

The Late Ordovician was also a time of serious volcanic activity. Although volcanoes usually raise global temperatures by pumping greenhouse gases such as carbon dioxide into the atmosphere, in this case these warming effects must have been offset by the cooling effect of the southern polar ice cap. So, it was another effect of volcanic eruptions that was crucial: the input of the element phosphorus into the oceans.

In 2021, Jack Longman, a geochemist then at the University of Oxford, and his colleagues, proposed that the drastic sea-level fall coupled with volcanic activity could have resulted in high levels of weathering of terrestrial rocks, including lavas and volcanic ashes, from which phosphorus was leached and washed into the sea. The excess phosphorus entering the oceans produced further cooling and led to anoxia (the absence of oxygen) and euxinia (excess sulphur) in the oceans, both of them killers (see pages 163–68). Phosphorus in the ocean gobbles up oxygen to produce phosphate ions, and further oxygen is lost because phosphates can lead to eutrophication – a process whereby water becomes progressively enriched with nutrients and increases the amount of plant and algae growth (bloom). Phosphate is a nutrient utilized by animals, so an excess can lead to a bloom, or massive overproduction of life, which then rapidly depletes oxygen in the water. This is a common phenomenon seen today when phosphate-rich fertilizers leach into rivers and lakes. It has often been noted that many examples of fossils from the crisis times of the Late Ordovician occur in black rocks with excess sulphur and phosphate, indicating anoxic conditions, something that occurred during all of the major mass extinctions.

Together, the fall in sea levels and increased weathering that led to anoxic ocean floors added to the stresses on life due to the dramatically cooler conditions caused by the southern polar ice cap. The greatest mystery yet to solve is why this Ordovician ice age was so short. We are currently in perhaps the later stages of an ice age, following the intense northern hemisphere freezing and snowy Pleistocene landscapes associated with woolly mammoths and Neanderthal peoples, but this present ice age is part of a longer-term cooling episode of tens of millions of years. The

Late Ordovician glaciation seems to have been a sharp event of as little as one million years. Before it, world climates were warm and balmy; after it, they returned to those conditions rather rapidly. It is not yet understood how such rapid shifts in global climatic conditions occur.

Although the first of the 'big five' mass extinctions may then remain quite mysterious, it was associated with some of the most profound changes in ocean life that we still witness in modern marine ecosystems. As the only mass extinction known in any detail that was driven by freezing (the much older Snowball Earth episode is discounted because we cannot say how that event generated extinctions), it stands on its own. It was followed by a rapidly changing world that witnessed new groups of animals, most notably a huge expansion in the types of fishes, as well as major steps in the conquest of land by plants and animals. Just as the Late Cambrian extinctions ushered in the three-dimensional world of the Ordovician, the Late Ordovician mass extinction was perhaps responsible for the new worlds of the Silurian and Devonian.

PART 2

Mid-Palaeozoic Events

444–252 million years ago

CHAPTER 4

The Move to Land and the Late Devonian Crisis

TERROR OF THE SEAS

If we spiral back in time to 365 million years ago in Ohio in the USA, we would be in a Late Devonian world that was very different from the Ordovician. We are 100 km (60 miles) offshore from a series of islands stretching across Ohio and Indiana, and north into Canada, the Cincinnati Arch, with ocean covering Michigan to the west and Ohio and Pennsylvania to the east. Looking into the depths, you cannot see the ocean floor – the waters are dark, mysterious and lacking oxygen. Indeed, the absence of oxygen (anoxia) means the seabed sediments are black and full of organic matter, although there is nothing living on the seabed to feed on the fallen plankton and fish carcasses.

A shoal of small, heavily scaled fishes, each 8 cm (3½ in.) long, flits past. These *Kentuckia*, with short heads covered in bony plates and a simple mouth that opens and shuts, slurp up plankton and plant fragments that have drifted in from the distant land. Their bodies bear sharp-pointed fins and a V-shaped tail, which they beat from side to side to propel themselves forward. Their bony, rectangular scales are arranged in orderly rows.

The shoal is pursued by a slender shark-like animal, 2 m (6½ ft) in length: *Cladoselache*. Although it looks like a modern mako shark, it is only very distantly related, despite its broad tail, large dorsal (back) fins and even larger pectoral (front) fins. Its head has a broad toothy mouth – an obvious predator – and its body is mainly black, with a white belly, a common type of countershading seen in fishes that live near the surface. A black back means they cannot be seen when viewed from above; they blend into the general blackness of the deep water. When seen from below, the white belly blends with the bright sunlight coming through the sea surface. The lazy *Cladoselache* swims among the much smaller *Kentuckia* shoal and scatters them to right and left by flicking its head. It then lunges and snaps, flipping its prey so the head points down its throat and then disappears.

A second *Cladoselache* exhibits the final fate of another prey fish by expelling a beautiful spiral faecal pellet from its anus. The pellet looks like an elongate pine cone, maybe 10 cm (4 in.) long, with eight turns of the screw and pointed ends; it twirls as it sinks, heavy with the undigested bony scales of the fish. Twenty seconds later, it hits the seabed and a flurry of black mud rises. Over time, the pellet will be covered with sediment and become invested with pyrite (an iron sulphide; see page 106) from the anoxic sediments. The pellet's pattern is formed by the spiral valve at the end of the fish's digestive system, which *Cladoselache* has in common with most present-day sharks. The spiral slows the final passage of the food waste to ensure that the maximum amount of nutrients can be extracted before the faeces are extruded; because *Cladoselache* were meat-eaters, these were often solid, compact structures, mainly gristle and bone.

A very different kind of fish moves slowly into view. It is 40 cm (16 in.) in length, with a head encased in a bony helmet made from many plates and a small bone-rimmed mouth near the front. Its body is long and the tail whip-like, flicking rapidly from side to side, as it moves fitfully, hunting prey. This is *Coccosteus*, a placoderm fish, a heavily armoured beast, first found in the 1830s in the Devonian rocks of northern Scotland, but also widespread and present in North America.

Suddenly, the scene appears to freeze. The armoured *Coccosteus* and the shark-like *Cladoselache* look about. The water darkens and all falls silent; the shoal of silvery *Kentuckia* has gone. The two *Cladoselache* flick their tails and disappear. So, too, does the lone *Coccosteus*. Something huge looms through the water. At first, a giant box-like head, 2 m (6½ ft) high and wide, can be seen, its two beady eyes flicking from left to right. Next, an array of massive jaw plates comes into view, spikes above and below, one on each side, and broad arcs of sharpened bone sweeping back round the huge grinning jaws. These are not the usual teeth of a predator, but great blades of bone, sharpened by cutting past each other, like the blades of a giant pair of shears.

The jaws open and shut slowly as the animal moves forward in a stately manner like a giant whale, although it is clearly not a plankton-eater; any unhappy fish that has not detected the silent arrival of this great predator will simply disappear inside its vast gullet. Its giant pectoral fins rise and fall gently as it steers to left and right. It has a 1-m (3-ft)-tall pointed dorsal fin and a tail that is 2 m (6½ ft) high and sweeps slowly from side to side, each movement shifting masses of water and providing forward pro-pulsion. Its body is covered with dark-coloured leathery skin, strengthened by numerous scales. This is *Dunkleosteus* and at nearly 9 m (29½ ft) long and 4 tonnes in weight, it is the largest animal yet on Earth, and certainly the largest fish of its day, and for a long time afterwards (see Plates IX, X).

Like *Coccosteus*, it is a placoderm, but much larger. Its jaws work on a complex two-hinged crank system, whereby they can snap shut in the usual way, but there is a second joint between the back of the head and the shoulder plates. In closing its jaws, it powers up with the lower jaw, as humans do, but can also swing the upper part of its head downwards with equivalent might. This generates 6,000 newtons (0.6 tonnes) of force at the tip of the jaws and 7,400 newtons (0.75 tonnes) of force further back in the jaws, at the centre of the cutting blades.

Three *Coccosteus* swim across in front of the behemoth. Unhurriedly, and without perturbing the water (the *Coccosteus*, like many fishes, have motion-sensing nervous systems around their heads), the giant *Dunkleosteus* lines up with its intended prey and moves forward. Its jaws open wide with a sudden crack and the *Coccosteus* disappear in a split second, deep into the throat. One of them frantically swims back out, pumping its tail as fast as it can, but as it passes over the cleaving blade on the right side, it is severed in two, its bony armour giving it as much protection as a can of beans under the metal tracks of an army tank. The *Dunkleosteus* crunches down and the prey disappear; it swims forward into the dark, unblinking.

SWIMMING WITH PLACODERMS

Dunkleosteus is best known from the Cleveland Shale of Ohio. This Late Devonian rock unit varies from 2 m (6½ ft) to 30 m (100 ft) thick around Cleveland, and it was once explored for its potential to produce oil and gas but was never as productive as the older Utica and Marcellus shales of this highly industrialized region. For palaeontologists, the Cleveland Shale is renowned for its fossils, especially the extraordinary diversity of the sixty-five fish species.

Particularly important are the placoderms. These heavily armoured fishes have been found in abundance in the Cleveland Shale. They were half bony armour over the head and shoulders, and half flexible body and tail behind, although even the hindquarters, used to drive themselves through the water by powerful lateral body bends, were covered with a tough chain mail of flexible bony scales. The head-and-shoulder shield was a complete box around the front end, composed of multiple curved bony plates.

The head section had wide shields over the top of the head and around the sides, a system of smaller plates around the nostrils and upper jaws, and even a bony headlight casing around the large eyes. The lower jaws were composed of single massive plates. The shoulder section was similarly made up of large bony plates, partly for protection of the vital internal organs and partly to support the front (pectoral) fins. Placoderms had large pectoral fins, but only small pelvic fins. The pectoral fins were used for steering the lumbering body as it ploughed through the water. By sticking a paddle out to the left, the body swerved to the right because the extended fin caused that side of the body to slow down. These front fins matter as we will later explore.

Museum visitors generally don't care about the fins, tails or scales of *Dunkleosteus*. They want to be photographed with their heads inside the metre-wide gape of the sharp-edged jaws. Close-up, those bevel-edged bony spikes and blades are truly terrifying, and when combined with the evident forces that could be generated, make a fiendish weapon (see Plates IX, X). The forces were calculated in detailed biomechanical reconstructions of the whole crank system of their jaws, with the two pivots, by US palaeotologists Philip Anderson and Mark Westneat.

But there was another effect: as *Dunkleosteus* opened its jaws sharply, prey shot into its gullet. This is a common adaptation in aquatic predators called suction feeding. Underwater, if an animal opens its jaws, the pressure inside reduces and water rushes in to fill the void. Usually, this rushing water also contains food, meaning that predators, such as fishes, underwater-feeding frogs and others, don't have to work too hard to be able to engulf their prey. (This effect is not replicated out of water because the density of air is much lower than water, and the force required to create a similar suction effect would be enormous.)

Dunkleosteus and the other fishes of the Cleveland Shale lived at a time of major environmental disturbance worldwide. Indeed, as we shall see, they were in the lull between two storms: a massive phase of extinction that had occurred ten million years earlier and another catastrophe that would happen shortly in the future. However, there had been a great deal of evolution on Earth since the Late Ordovician mass extinction, and most notably, life had moved definitively onto land. The armoured fishes of the Devonian played their part in this process.

CREEPING ONTO LAND

Going back in time from the Cleveland Shale, it is generally assumed that there was not much life on land until the Silurian Period, beginning 444 million years ago. However, it's impossible to say when the first microbes became adapted to living out of water. This is especially true of photosynthesizing cyanobacteria and algae, which had to live near the surface of the water in order to be able to capture energy from sunlight. As today, some of these green slimes and seaweed-like plants most likely occupied coasts and shorelines and would have been exposed to the air by the tides, although it's hard to know how many of these might have lived most of their lives out of the water. Other microbes and fungi must also have been constantly in and out of the shallow waters.

The first definite evidence of life on land consists of some Ordovician-aged soils. These are preserved as blocks of sediment, with mysterious meandering tubes penetrating through the blocks. The tubes may have been formed either by the roots of some plants or by small burrowing animals. Either way, soils are generally formed on land and so indicate life on land: a soil is simply ground rock plus organic matter, but the organic matter itself is generated either from rotting plant material, plus animal bits and pieces, or by the reworking of animals, such as earthworms that chomp their way through the soil and aerate it. There are also some rare, but convincing, fossils that confirm there was plant life on land in the Ordovician. These are microscopic spores of bryophytes, the simple green plants we now recognize as mosses and liverworts. Their distant ancestors were already living in damp spots around ponds and shady rocks and slowly building small thicknesses of soil. These soils then provided living opportunities for the more complex plants that were to come in the Silurian.

The most famous of the Silurian plants is *Cooksonia* (see Plate VII), named in 1937 after Australian palaeobotanist Isabel Cookson (1893–1973), who made great contributions to the understanding of these early days of plants. At a Silurian lakeside, from a distance, you might have seen a green sward for several metres around the waterside, but close up you would see short 6-cm (2½-in.)-long stems standing vertical, splitting near the top into two or four little branches, each tipped by a button-shaped blob. These terminal blobs were the spore cases, which grew and swelled, and at the appropriate time probably split open, scattering the spores, which fell to the ground and produced new plants.

Even though these fields of *Cooksonia* were modest, the plant had an important feature that was to enable it and the other early tracheophyte plants – essentially, all the green plants except algae and mosses – to conquer the world: their stems were stiffened by internal tubes (tracheids) that conducted water from the ground up to the tips. The tracheids were reinforced by beautiful strips and spirals of lignin, a complex biomolecule that allowed plants to grow upwards (lignin is what makes wood woody because it is stiff and slow to rot).

The Silurian plants were tiny and kept close to water, but as soils built up, generated by the plants themselves, in the Devonian (starting 419 million years ago), tracheophyte plants became taller, to perhaps 50 cm (1–2 ft), and had tangled root-like structures connecting them all together and rooting them in the soil, but also dipping into the freshwater of lakes or rivers to obtain water. By the end of the Devonian, when *Dunkleosteus* and contemporaries were cruising the Ohio seas, some of the primitive tracheophytes had become bushlike, and some treelike. The extraordinary *Archaeopteris* even reached the amazing height of 24 m (80 ft), with a massive trunk composed of huge numbers of lignin-strengthened tracheids that gave it support. Further, these huge tree trunks shaded and protected an undergrowth of smaller plants on the ground, lycophytes, horsetails and ferns.

In fact, because I am not very car proud, I see some of this evolution occurring today, in the corners of my car windows. First, as dust gathers on the glass, it is washed by rain into the rubber corners of the windows, where after a few months, tiny bryophytes take root. These mosses and liverworts are adapted to grow on very little soil, often colonizing rocks or dead timber. Then, after a year of moss growth in the corner of the car window, I've noticed small tufts of grass, just a few millimetres tall, taking root in the soil made by the bryophytes. Such is the amazing tenacity of plants. In the Silurian, these primitive plants provided opportunities for a new invasion of the land.

ANIMALS MOVE ONTO LAND

In the same way that it is not known in detail when all early plants conquered the land, the record of early land animals is also patchy. We do know, however, that the ancestors of insects, spiders and millipedes had already made the move by the end of the Silurian. Not only did they have to switch from breathing underwater to extracting oxygen directly from the air, they also had to support their weight and protect themselves from drying out. In the water, animals essentially weigh nothing and move about by swimming gracefully. On land, they have to hold up their bodies, meaning the whole structure has to change to support the internal organs. Today, beached whales and dolphins can die from the crushing of their internal organs under gravity: even though they breathe air like other mammals, these marine creatures have adapted to the neutral buoyancy of living in water and their internal organs, such as lungs, heart and guts, float in their correct places. On land, their organs collapse.

The Silurian arthropods that first crept out on to land were much smaller than present-day whales and dolphins, which made gravity less of a problem for them because, while the material strength of cells and tissues of the body remains constant regardless of size, the effect of gravity was much less than in a large animal. Fossils of these early centipedes, spiders and insects are known because their external skeletons were tough and have been fossilized; there are also fossils of early land snails and other molluscs that made the move onto land. There must also have been ancestral earthworms and round worms moving in the soil, but their fossils are absent. Where such tiny creatures trod, larger beasts were bound to follow.

Fishes made the transition to living on land in a series of stages during the Devonian. A key to this was the famous fin-to-limb transition, which happened in a stepwise manner. Fishes have paired pectoral and pelvic fins, and, as we have seen, versions of these are seen in the placoderms, as well as in all other fishes, including those early sharks like *Cladoselache* and the bony fishes of the Devonian such as *Kentuckia*. Fins are the equivalent of our arms in terms of evolution. We have two legs and two arms, and these four members give the name 'tetrapod' ('four foot') to all the vertebrates that walk on land – the modern amphibians, reptiles, birds and mammals. Technically, our arms are pectoral appendages, and our legs are pelvic appendages. They are attached to the body through the pectoral and pelvic girdles – our shoulders and hips.

Palaeontologists are keen to understand the fin-to-limb transition – the way in which tetrapods emerged from the water at some point during the Devonian and started walking on limbs that were weight-bearing and moved in an episodic, back-and-forwards manner to enable an animal to progress, rather than swimming with their whole bodies and steering with fins that did not support any weight. In the past hundred years, the search has been for Devonian fishes and early tetrapods that document this transition, and many examples have been found. Critical were the discoveries of two tetrapods in the Late Devonian sediments of Greenland. The field campaigns by Danish geologists in the 1920s were heroic, as they rushed to the frozen shores of eastern Greenland in the short summer when they could hack out the rocks; in the Devonian, these areas experienced warm and humid climates as they lay much closer to the Equator. What excited the palaeontologists was that the tetrapods *Ichthyostega* and *Acanthostega* seemed to be halfway between fish and tetrapods. They had definite arms and legs, with all the key bones, but they also had torpedo-shaped bodies and tails with long fins, so they still spent time swimming in lakes seeking fishy prey.

Other early tetrapods subsequently found in Europe, North America and China show many further modifications to the limbs, and a number of remarkable fishes from the Middle and Late Devonian such as *Eusthenopteron* from Scotland and Canada, and *Tiktaalik* from Arctic Canada, help fill the gaps. *Eusthenopteron* was one of an unusual group of bony fishes related to modern lungfish, and definitely very much a fish. It was about 1.5 m (5 ft) long, with unusual muscular pectoral and pelvic fins, which it used to drag itself around on the bottoms of lakes. The fins had not only thin bony rays, but also some key limb bones, such as the humerus (upper forelimb) in the pectoral fin and the femur (thigh bone) in the pelvic fin. *Tiktaalik*, even larger at 2.5 m (8 ft) long, had more complex pectoral fins capable of bending and some form of walking, although this was with its belly on the ground. Its head was broad and streamlined, and it probably spent most of its time in ponds. It has even been suggested that because climates were warm, in summer these ponds might have dried up, so *Eusthenopteron* and *Tiktaalik* had to drag themselves over squelchy mud looking for protection in another pond. Breathing air was not an enormous problem for these fishapods (half fish, half tetrapod), as some have called them, because they all had lungs and could breathe air while relaxing in a warm pond or could filter oxygen from the water through their gills.

At some point, these occasional visitors to the land must have snapped at a fat spider or millipede, or tried a mouthful of plant stems near the water's edge, adding a useful supplement to their usual, fishy diet. The transition was underway, and these early, tentative steps on land marked the beginning of the land vertebrates. All this development, however, was taking place in a world that was heading for two disasters, and these crises neatly bookended the span of time when *Dunkleosteus* and friends were enjoying the rich living of the Cleveland Shale.

THE KELLWASSER AND HANGENBERG BOOKENDS

The Late Devonian mass extinction divides into two events, separated by 13 million years and bookending the Famennian Stage, the final division of the Devonian, which lasted from 372.2 to 358.9 million years ago. It was named in 1855 after the Famenne region in southern Belgium, where rocks of the time are abundant, equivalent in age to the Cleveland Shale of Ohio and that even share some of the same kinds of placoderms and other fishes. The two Late Devonian extinctions are named after their classic representative rock sections in southern Germany: the first, at Kellwasser near Hanover in Lower Saxony, and the second at Hangenberg in the Rhenish Massif.

The Late Devonian was a time of considerable environmental upheaval, with sea levels changing from high to low and many phases of anoxia on the seabed, as was seen to affect the Cleveland Shale. There were also rapid changes in temperature. However, the complete picture for the entire Famennian Stage, as well as the particular crises at the start and end, has been hard to establish.

Massive volcanic eruptions were a plausible cause of the Kellwasser event about 372 million years ago. The Viluy Traps are thick accumulations of basalt, a black-coloured rock formed from molten magma, which cover hundreds of thousands of square kilometres of northern Siberia and are estimated to have poured out more than 1 million cu. km (240,000 cu. miles) of lava. The date of maximum eruption of these lavas coincides with the date of the Kellwasser mass extinction and this might be informative. Although life forms around an active volcano are killed by lava, it is not only the lava that kills life. Accompanying the outpourings of molten rocks are huge volumes of greenhouse gases such as carbon dioxide, methane and water vapour, which are pumped into the atmosphere. These cause warming over a wide area, even worldwide, and this can be associated with acid rain on land, acidification of seawater and anoxia on the seabed, often associated with high sulphur levels, as we will explore further in Chapter 6.

Geologists have also sought evidence for similar, large-scale volcanic eruptions at the time of the Hangenberg crisis about 359 million years ago, but have so far not found a suitable smoking gun. Instead, an unusual mechanism has been identified for this event, which was perhaps driven by the success of plants in colonizing the land throughout the Devonian: the reduction in the amount of carbon dioxide in the atmosphere. Plants absorb carbon dioxide as they photosynthesize, and a lowering of carbon dioxide concentration leads to global cooling. There is indeed evidence for another southern polar glaciation at this time, perhaps not so severe as in the Late Ordovician, but glaciation nonetheless. Again, as well as a general worldwide cooling, the developing ice cap drew water from the oceans so sea levels also fell.

The land plants might have also added to the problems caused by lower sea levels. As we saw in the Late Ordovician crisis (see Chapter 3), lowered sea levels exposed former seabed to the atmosphere, which reduced space for all the rich life of the marine continental shelf. The new plants sent roots into soils and especially rock surfaces, extracting minerals for their own growth, but this new activity also increased overall rates of rock weathering and so nutritious minerals such as phosphorus were washed into streams, lakes and the sea. These minerals might have led to eutrophication as in the Late Ordovician (see page 65), leading to anoxia in seabed sediments.

Whatever the mechanisms that caused the crisis, its scale is not in doubt. The Kellwasser and Hangenberg events accounted for losses of more than 50% of species in the oceans and on land, with the rates of loss appearing to have been higher during the Hangenberg event. The effects were clear in terms of ecosystems. Most of the fishes in the Cleveland Shale disappeared, including all of the placoderms. Through the Devonian, these heavily armoured fishes were not all monsters like *Dunkleosteus*, but included hundreds of other species that were an important part of every marine or freshwater environment. They were mainly carnivores, but smaller species may well have fed on smaller food elements. Numerous other Devonian fishes disappeared, many of them armoured, although survivors included the bony fishes like *Kentuckia* and the sharky predators like *Cladoselache*.

Extinctions of other marine life were equally abrupt and shocking. Whole plankton groups disappeared and there were also losses among reef-building corals and sponges. This was serious, as many areas in the tropics had seen the development of enormous and diverse reefs during the Devonian. They completely disappeared, with a reef gap of some ten million years following the Hangenberg event, meaning reefs took that span of time to recover. There were also extinctions among trilobites and molluscs. Also gone were many of the land plants and animals, including the strange Late Devonian trees such as *Archaeopteris* and all the fishapods, like *Tiktaalik*, *Acanthostega* and *Ichthyostega*.

The Late Devonian mass extinction has long been tagged as the second of the 'big five'. However, these designations are not always helpful: as we have seen, there were several mass extinctions before the 'first' Late Ordovician crisis, and the Late Devonian mass extinction was not a single disaster, but at least two, separated by a long span of time. What is more important than terminology, however, is the observation that crises opened up opportunities for new life, and this also applied to the Late Devonian. Having cleared away reefs, many armoured fishes and much of land life, the rise of some entirely new ecosystems in the succeeding Carboniferous and Permian periods was triggered. These were times of warm worlds – humid and then arid – and life flourished. A seeming anomaly, however, is that while warm temperatures can encourage richness of life, they can also kill: in the next chapter, we explore how warm temperatures can both nurture and diminish biodiversity.

CHAPTER 5

How Global Warming Kills

GIANT FERNS AND DRAGONFLIES

During the history of the Earth, high temperatures have occurred many times. In terms of ecology, heat can be good, but heat also can be bad (and a cause of many mass extinctions). In this chapter, we will disentangle the benefits and dangers of warm temperatures as they occurred in the Carboniferous and Permian, and also pinpoint the mechanisms by which global warming kills.

The classic scene of the Late Carboniferous in many parts of the world, from 323 to 305 million years ago, included tall seed ferns and horsetails, green and dripping with moisture, and streams and ponds filled with a variety of water-loving tetrapods and fishes. Giant insects and spiders flew and scuttled in among the trees, such as the dragonfly *Meganeura* that reached wingspans of 70 cm (28 in.), making it more seagull than insect, and the millipede *Arthropleura*, at 2.5 m (8 ft) long, the length of a car. As the creature ploughed through the plant debris on the forest floor, it hoovered up vast quantities like an industrial vacuum cleaner.

Deeper in the freshwaters around the forests, new kinds of fishes patrolled the waters. Gone were the placoderms and other

Meganeura, a giant dragonfly as large as a seagull, flew through the tropical forests of the Late Carboniferous.

armoured forms of the Devonian, and in came shark-like fishes. Indeed, some were very strange, with outgrowths of bone over their heads, some like an umbrella handle and others like a woodworker's rasp file covered in teeth on the top surface. What function these structures served is not known – but perhaps they were for sexual display, like the elaborate tails of some male birds. The bony fishes also survived from the Devonian and became diverse, looking like modern salmon or cod, but with short, bony-plated heads and heavy bony plates instead of thin scales over their flanks. Unlike the complex jaw system of modern cod and salmon, however, theirs was simple – they could only open and shut their jaws, like the lid of a kitchen refuse bin.

This was the new world that was enabled or triggered following the Late Devonian mass extinctions. The dramatic losses of life in the Kellwasser and Hangenberg events opened up opportunities for many new groups of corals, brachiopods, trilobites and other sea beasts, as well as the remarkable new groups of fishes, plants and tetrapods.

The zones of rich life in the Carboniferous lay around the Equator of the day. Likewise, in humid tropical zones in the present day, high temperatures and abundant supplies of water provide the basis for very high biodiversity. We are familiar with images of Brazilian or Indonesian rainforests where trees are covered with ferns and lianas and are abuzz with insects, colourful birds, monkeys and all kinds of exotic life. In continent-scale comparisons, ecologists find very strong relationships between temperature and biodiversity, and rainfall and biodiversity. For example, in the ice-covered parts of North America, ornithologists struggle to identify four or five species of birds, whereas in the tropics of Central America their checklists may extend to over five hundred species. This remarkable biodiversity gradient from Pole to Equator is an example of what biogeographers call the 'latitudinal diversity gradient'. In this case, the species diversity of birds increases one-hundredfold from northern Alaska to Guatemala. This is the same with the presence of water. From deserts (cold, as in southern regions of Chile or Argentina, as well as hot) to humid zones, the numbers of plant, insect or vertebrate species increases from tens to hundreds or thousands. The main driver of biodiversity on land seems to be temperature, although water is essential too; as we shall see, the balance of high temperatures and limited water supplies can suddenly become fatal.

CARBONIFEROUS RAINFOREST COLLAPSE

The interplay of temperature and humidity were crucial in driving some important events in the evolution of life through the Carboniferous and subsequent Permian periods. On land, the rich, tropical forests of the Late Carboniferous extended over modern North America and Europe, which at the time formed a single great land mass astride the Equator. Since the Ordovician, most continents had moved northwards, drawing much of Gondwana away from the South Pole and shifting many of them into the tropical belt.

It's thanks to these great forests (see Plate VIII), full of giant, exotic trees, bird-sized insects and early tetrapods, that many nations in Europe and North America are blessed with huge coal reserves (or, at least, they thought they were blessed at the time of the industrial revolution in the 1700s to early 1900s when coal drove the furnaces of industrial plants and the new, expanding cities). Nearly all the coal used in North America and Europe, from England to Poland, is Late Carboniferous. Now, of course, most people see coal as a danger because burning it releases carbon dioxide, a greenhouse gas that causes global warming (we'll come back to this in Chapter 15).

The lush Late Carboniferous coal forests did not last, however. They collapsed quite rapidly about 305 million years ago, although not everywhere – for example, they persisted in much of what is now China, where the main coal reserves are Late Permian in age. A 2010 paper led by palaeontology PhD student Sarda Sahney, with palaeobotanist and palaeoclimate expert Howard Falcon-Lang, both then at the University of Bristol, explored the rapidity of the collapse in North America and Europe and its apparent connection with a major fall in sea level of about 100 m (330 ft).

The model for this collapse is similar to that of the Late Ordovician mass extinction discussed in Chapter 3 – the main cause being the southern polar ice cap, which drove a fall in sea levels and led to cooler and drier climates. A large ice cap had developed over the South Pole near the end of the Carboniferous, which lasted well into the Permian. A huge sea-level fall followed because so much water was locked into the ice, and cool, dry climates certainly developed over much of the tropical European–North American land mass. There were also major volcanic eruptions at this time, from the Skagerrak-Centred Large Igneous Province, a grand title for a large and scary series of volcanic eruptions located around Sweden, Norway and the North Sea. Could these eruptions have contributed to the general disaster? More research is needed to answer this, but we do know for certain that a new kind of hyperthermal killing model involving sharp temperature increases caused a number of disasters in the Permian.

THE HOT PERMIAN

Fast forward 15 million years to the Nocona Formation in Texas, USA, where a famous fossil bed called the Geraldine Bonebed has produced thousands of specimens, including plants, insects, fishes, amphibians and reptiles. Named after a nearby ghost town, the bone bed was discovered by the great American vertebrate palae-ontologist Al Romer in 1932. The sediments and plants found suggested that the area was a lushly vegetated floodplain with small lakes and a swamp forest of ferns, conifer trees and seed ferns.

As we enter this scene, the most impressive and by far the largest animal of its day we see is *Dimetrodon* (some reached 4 m/13 ft in length), which walks on all fours and has a chunky head and curved jaws armed with sharp predatory teeth. Most extraordinary is the sail along its back, a huge structure supported by long bony

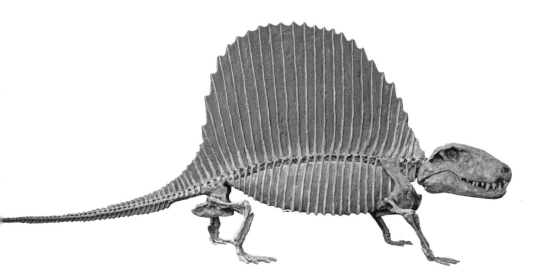

Skeleton of *Dimetrodon* showing its low-slung body, large head and toothy, flesh-eating jaws, as well as its remarkable dorsal crest, made from skin stretched over the elongate neural spines of the vertebrae of the back.

extensions from each vertebra of the backbone and extending from the back of the head to the beginning of the tail. The sail was probably made from thick layers of skin that were slung between the vertical bony supports; here and there it is torn and ragged thanks to rough encounters with other Dimetrodon or spiny plants.

The Dimetrodon stalks along looking for prey and spies a couple of 3.5-m (11-ft-6-in.)-long Edaphosaurus. These reptiles also carry sails on their backs, but have much smaller heads and small teeth adapted for feeding on plants. Their fatter bodies accommodate the very large gut required by herbivores to enable the digestion of the rather poor plant food available. The Edaphosaurus are serenely chomping great mouthfuls of horsetails they have harvested from the edge of the pond, seemingly unaware of any danger. But, as the Dimetrodon steps through the scrub, one of the bulky herbivores hears a twig crack and chirps sharply at its mate, and they waddle off. These two are nearly as big as the Dimetrodon, so it decides to await a smaller prey animal to save its energy.

Dimetrodon and Edaphosaurus are synapsids, members of the large tetrapod group that later gave rise to mammals. The functions of their sails have long been debated; the most commonly held view is that they were thermoregulatory structures. Perhaps, so the argument goes, these early synapsids were cold-blooded, like modern lizards, and would have been torpid in the cold early morning air. Today, lizards may shut down and doze during the cold night, then come out and sprawl on rocks in the early morning sun, basking to absorb warmth. Likewise, the sail-backed synapsids may have done the same, acquiring heat from the Sun through the sail, which contained rich networks of blood vessels allowing the heat to pass to the body. This could have given them an advantage in getting moving at daybreak. It is also argued that in the heat of the Early Permian midday, the sailbacks could have stood in the shade and radiated heat from the sails in order to cool down.

Whether this model is true or not cannot be proven. An argument against it is that the other reptiles of the Geraldine Bonebed lacked such sails, and yet they, too, must have coped with the daily temperature changes from day to night and avoided attack by the sharper *Dimetrodon*. On the other hand, *Dimetrodon* and *Edaphosaurus* were bigger than the other coeval animals, so perhaps needed this boost.

Synapsids seem to have lived fairly worldwide, with similar examples also known from Germany. There was no Atlantic Ocean at this time; the North American–European supercontinent that had existed in the Carboniferous, and was the site of all the coal beds, persisted into the Permian, when it was joined by the northwards-drifting southern continents, collectively called Gondwana. During the Permian and subsequent Triassic periods, the world was essentially a single supercontinent called Pangaea (meaning 'all world'). Climates were hot throughout, and dry in the huge desert-like interior of the supercontinent.

About 273 million years ago, at the boundary between the Early and Middle Permian, the age of *Dimetrodon* and *Edaphosaurus* came to an end, possibly as a result of this extreme, hot climate. The causes of the extinctions of tetrapods remain uncertain, but their disappearance led to the emergence of new reptile groups and the beginnings of some major changes in their modes of life.

THE END-CAPITANIAN MASS EXTINCTION

We move from North America and Germany to Russia and South Africa to understand the Middle Permian reptiles. These geographic shifts simply reflect where the best fossil records are. In North America and Europe, the red-coloured desert sandstones were flooded by shallow seas and so fossils of land-living reptiles are not seen. In Russia, on the other hand, what had been under the shallow sea became land, and in South Africa, areas covered by the retreating southern polar ice sheet became habitable again and land plants, insects and tetrapods returned.

A great deal of extinction had been noted around the end of the Middle Permian, at the end of the Capitanian Stage about

259 million years ago, but rock dating was uncertain. It was particularly difficult to compare ages from the rock successions in South Africa to those in Russia and China, and especially to match rocks deposited on land with those in the sea. But thanks to detailed studies in South Africa by palaeontologists Michael Day and Bruce Rubidge, we can now be sure that there was a big extinction event at this time.

On land, Day and Rubidge noted a major changeover in the reptiles at this point. Two typical groups in the Middle Permian were the bradysaurs and the dinocephalians. *Bradysaurus* was a pareiasaur, a unique group of reptiles known only from the Permian. Hefty creatures, about 2.5 m (8 ft) long and weighing up to half a tonne (1,100 lb), they were as bulky as a hippopotamus but with a tiny head, and had armour plates over their backs. The arms and legs were relatively short and stocky, and the animal waddled along, with its limbs sprawling out to the side. With its short teeth and massive jaw muscles, it chomped through tough plants such as ferns and horsetails and probably required a bulky body to enable it to digest this poorly nutritious food.

The dinocephalians were synapsids, descendants of *Dimetrodon*, and include both herbivores and carnivores. One of the typical herbivorous forms, *Tapinocephalus*, was even larger than *Bradysaurus*, measuring 3 m (10 ft) long and weighing 1–1.5 tonnes. It had short, sprawling arms and legs, and the body sloped up from a low hip region to powerful shoulders that held its head high. The cranium was topped by a thick-boned bump on top – these big beasts might well have crashed heads together in mating contests, just as mountain sheep do today. The bradysaurs, dinocephalians and many other tetrapods died out at the end of the Capitanian. In the oceans, there were also many losses of brachiopod and ammonite species, and a substantial restructuring of reef ecosystems. Major planktonic organisms disappeared. The same extinctions are seen in rocks of the same age in South Africa, Russia and China. But what caused them?

In the oceans, there is evidence that sea levels fell and that sediments on the ocean floor experienced a time of poor oxygen content (anoxia), which can kill seabed life. There are also suggestions of ocean acidification, which, if it happened, would have attacked the limy skeletons and shells of corals, brachiopods and other marine life. In the red sandstones in the great Karoo basin of South Africa, Day and Rubidge noted evidence for drying climates and that plants had been dying. As the plants died back, normal processes of erosion washed away the soils and stripped the landscapes back to bare rock.

It has been suggested that these physical changes in the oceans and on the land were driven by huge volcanic eruptions in what is now South China. The so-called Emeishan volcanic rocks are the right age, and eruptions seem to have peaked just at the end of the Capitanian. As we will explore further in the next chapter, volcanoes not only kill life in the close vicinity from the molten lava and hot rocks thrown in the air, but also through the gases that are pumped into the atmosphere. These volcanic gases would have mixed with rainwater to produce acid rain, which killed plants on land as well as acidified the shallow ocean waters. Most significant would have been carbon dioxide, a well-known greenhouse gas, that might have raised temperatures in the atmosphere, although evidence for this is elusive. There is evidence of temperature rises in Chinese rock sections just before the end-Capitanian event, and these may have been significant in driving the initial stages of the extinction. This mass extinction was followed by a time when tetrapods existed at low diversity; eventually, new species plugged the gaps in the ecosystems and life returned to its full richness.

PROFESSOR VERNON EXPERIMENTS

It might seem evident that animals and plants die at high temperatures, but exactly how high? This question matters because we are concerned today about global warming, and about its effects on biodiversity as well as on human health. British physiologist Professor Horace Middleton Vernon (1870–1951) tackled both topics, one pretty ghoulish, but the other with the most humane of motives. Vernon trained as a medical doctor as well as a zoologist. He spent time at the famous Naples Biological Station in Italy, where generations of marine biologists have been trained, and then became University Lecturer in Chemical Physiology at the University of Oxford.

While at Naples in 1899, he set about boiling a whole variety of marine animals until they died to determine their average death temperatures. Working his way through twenty species of corals, molluscs and vertebrates, he found these were in the range between 34 and 40.6°C (93–105.1°F). He described his procedure: 'The method of experiment was to place the animal under observation in a small beaker of water … This water was gradually warmed, at first fairly rapidly, but as the death temperature of the animal was reached, very slowly. The water in the small beaker was stirred vigorously with a thermometer, the animal being observed at short intervals so as to determine exactly when heat paralysis set in, and all motion ceased.' The water was then cooled, and if the animal recovered and started to move about, it was heated up to a temperature one degree higher than previously. The heating and cooling cycles were repeated until the poor creature had indeed been boiled to death.

Vernon established that, on average, fishes die at around 39°C (102°F), amphibians at 39.5°C (103°F), reptiles at 45°C (113°F) and various molluscs at 46°C (115°F). Overall, he noted that the death temperature seemed to depend on the balance between fluid and solid portions of the anatomy. Marine organisms that are little more than bags of water with a few solid organs died first, whereas vertebrates, for example, with their much more muscular and solid anatomies, could survive to higher temperatures.

This all seems rather ghoulish and downright cruel, but the facts were established and the experiments do not have to be repeated. Much of Vernon's other work on the effects of heat on life were performed on dissected muscle tissues, not on living animals. During the First World War, Vernon was engaged in gruelling factory work making bombs and observed that after long work shifts, often at high temperatures, workers were distressed and liable to make dangerous mistakes. He began to write a series of reports and books about this, and after the war, gave up his Oxford positions and focused on the new field of occupational health, campaigning for fairer working conditions, including laws to regulate temperatures in factories and other places of work.

The basic data established by Vernon and fellow researchers in the days when experiments on live animals were considered acceptable are surprising. The death temperature range of 34–40°C (93–104°F) is what we experience every day when we have a hot shower (at 37–38.3°C/98–101°F). Also, when we consider that temperatures of more than 40°C (104°F) were reported in southern Europe, the southern United States and China in the heatwaves of the early 2020s, it's evident that most of life would not be at all comfortable in these conditions. More significantly, modern physiological research shows that air and water temperatures above 28°C (82°F) are all in the 'stressful range' for marine life. In a 2022 study, physiologist Lisa Bjerregaard Jørgensen and colleagues from Aarhus University in Denmark showed that for each degree Celsius of temperature increase in the stressful range, the heat failure rate of chemical processes in the body, such as energy transfer, heart rate and oxygen consumption, increases by 100%. Cold-blooded animals such as molluscs, crustaceans, insects, fishes and frogs struggle and die. As Jørgensen and colleagues conclude, 'both aquatic and terrestrial ectotherms risk considerable increases in heat stress with global warming and this increase will be accentuated markedly on the regional scale and with each degree of further global warming'.

HOW HEAT KILLS

High-temperature events are called hyperthermals, and geologists have identified dozens of them, ranging in scale from the huge end-Permian mass extinction (see Chapter 6), middle-sized mass extinctions, such as the end-Capitanian event and two mass extinctions in the Late Triassic (see Chapter 9), as well as smaller-scale extinctions such as the Toarcian and Cretaceous oceanic anoxic events (see Chapter 10). It may seem inexplicable that high temperature can be such a killer – if, for example, temperatures rise by only a few degrees, as they did in the Capitanian, wouldn't plants and animals somehow just adapt? How does heat kill?

Research into heat stress emphasizes it has multiple effects. For example, as air temperatures rise, animals on land seek the shade: humans and cattle disappear under the shade of a tree, snakes and lizards squeeze under rocks or burrow into the soil. As temperatures rise further, animals start to shed water to cool down. Humans and horses sweat, where water oozes from pores all over their bodies; dogs and other animals pant. As temperatures rise ever higher, the water loss through breathing can increase dramatically. These linked responses are particularly dangerous. Effectively, an animal suffering heat stress needs additional water in order to be able to cool down, but high temperatures on land are often associated with a shortage of water, as, for example, in overheated tropical zones in summer.

Animals that live around deserts have adaptations to survive drying conditions. For example, camels have nostril flaps to limit water loss when they breathe out. Insects and spiders have cuticles enclosing their bodies that contain waxy compounds, and they can close small breathing tubes to keep water vapour in. There are some incredible survivors, such as the chironomid midge *Polypedilum*, which can tolerate an amazing range of temperatures, from −270 to +106°C (−454–+223°F). When temperatures edge towards the extremes, the midges pass into a dormant phase when most of their body systems shut down; examples are known that have recovered after seventeen years of dehydration and shutdown. Survivors such as these, however, are the exception. As Vernon showed, most animals die if temperatures reach 34–40°C (93–104°F), and most animals have no particular adaptations to survive the gruelling conditions of excess heat and aridity found in deserts.

Most land plants also die at these kinds of temperatures. Several things happen: their leaves burn up and fall off, and heat stress can reduce their normal processes of photosynthesis and even reverse the process to photorespiration, where they do not produce oxygen but absorb it, and their normal growth processes fail. Some simple plants such as mosses and liverworts can survive desiccation, but vascular plants such as ferns, conifers and flowering plants rely on water transport to extract minerals from the soil and to keep their stems and leaves healthy. During prolonged droughts, most plants go brown and eventually die.

There are some 'resurrection plants' that can recover from severe drying. Their cell membranes show special adaptations to protect them from water loss, and they have desiccation-induced protective proteins. In a time-lapse film on YouTube on the Boxlapse channel, the presenter places a Rose of Jericho in a bowl of water. The scrunched up, dry, grey-coloured plant slowly opens out, extending its long, seaweed-like leaves and stems, revealing healthy deep green shoots in the middle. The branches spread out further and the whole plant goes dark green in the course of an hour. But not many plants can do this – only 300 out of all 300,000 species of flowering plants.

High temperatures also kill in the oceans, as Vernon showed, and the critical death temperatures are the same. Modern researchers have found something unexpected, however – the critical temperature can be raised if the temperature rises fast, whereas under slowly increasing temperatures, species gets more distressed. Could this mean that sea creatures might survive the fast global warming associated with some mass extinction crises? The

The Rose of Jericho 'resurrection plant' that curls up when under severe heat stress (left) but opens up (right) when conditions improve.

answer is no, because the ability to survive a rapid rise in temperature is a short-term adaptation for sharp heat shocks that stop and reverse quickly. The animal switches on an emergency metabolic system, just as long-distance runners may switch from aerobic to anaerobic metabolism at the end of a race. When their oxygen supply is inadequate, the runner can burn sugars to power their muscles anaerobically, but this can give them cramps and the body has to recover.

A further discovery is that marine animals may be affected by multiple stressors when temperatures rise. As the water warms up, many marine animals effectively gasp for breath and need more oxygen (we do the same if we try to exert ourselves in a very hot sauna or hot pool), but there is no more oxygen. This is true of humans and farm animals, too, where hot stuffy conditions massively reduce wellbeing due to the combination of heat and low oxygen levels. We've come back to Horace Vernon and his explanation about why human beings require well-aerated workplaces and regulated temperatures. Heat kills.

Experiments on heat stress in modern plants and animals come together with palaeontology when we examine many of the later mass extinctions. Whereas some earlier events such as the Late Cambrian, Late Ordovician and Late Devonian mass extinctions might have been driven by climatic cooling, others, such as the end-Capitanian event, seem to have involved a phase of extreme heating caused by volcanic eruptions. This is also true of the mass extinction at the end of the Permian, the event that nearly terminated all of life and the subject of the next chapter.

Ⅰ Apocalyptic post-extinction scene. A small herd of *Lystrosaurus* hurry away from the erupting lava pools formed from the massive Siberian Traps eruptions at the end of the Permian, 252 million years ago.

Ⅱ The creativity of evolution. In the world after mass extinctions, new species evolve, and whole new modes of life can emerge. Here the small feathered dinosaur *Velociraptor* chases two small mammals of the genus *Zalambdalestes* in the Early Cretaceous of China.

III The first animals. Seabed scene over 555 million years ago, showing vertical frond-like creatures that fed by filtering small particles from the water. More mobile animals, some maybe relatives of worms or molluscs, creep around in the mud.

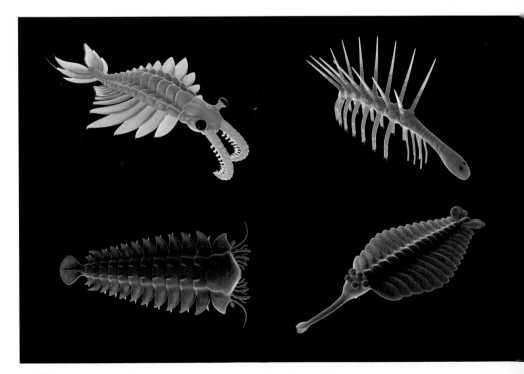

IV The wonders of the Burgess Shale. The Cambrian animals of this famous locality in Canada have never ceased to amaze. Here, we see, from top left clockwise, *Anomalocaris*, *Hallucigenia*, *Opabinia* and *Sanctacaris*, the latter an early arthropod related to modern scorpions.

(**v**) Trilobites. Here, a single layer of rock from the Middle Cambrian shows numerous examples of the trilobite *Ellipsocephalus hoffi*. These arthropods had walking legs beneath the carapace and ploughed the seabed mud for food; they could also scuttle about rapidly to escape predation.

(**VI**) Life in the Burgess Shale. This scene is crowded with examples of some of the weirder Burgess animals, as well as trilobites (left, bottom right), jellyfish (top left), and simple sponges (middle). The early vertebrate *Pikaia*, similar to *Haikouichthys*, swims across the centre.

VII Earliest land plants. Worm's-eye view of the tiny vascular plants *Cooksonia* that began to flourish in the Middle Silurian, 430 million years ago. Short, branching stems rise a few centimetres from connecting underground roots, and bulb-like spore cases top each stem.

VIII The great coal forests. Warm, damp climates prevailed over much of the equatorial lands in the Carboniferous 315 million years ago, and massive seed fern and horsetail trees flourished. Insects and other bugs flourished too, providing abundant food for early tetrapods.

(IX) The first giant predator. *Dunkleosteus*, the first truly giant predatory vertebrate, at 9 m (29 ft) long, prowled the Late Devonian seas of North America over 370 million years ago, snapping up every other fish it could find, even armoured specimens.

(X) Maximizing hunting efficiency. The *Dunkleosteus* jaws were powered by two jaw joints, mainly the usual hinge joint at the back of the mouth. But the front part of the skull also tilted at a ball-and-socket joint in the neck, providing extra power for the bite.

XI Munching machine. The humble rhynchosaur *Hyperodapedon* of the Triassic typified a new kind of herbivore. Abundant examples worldwide dominated herbivore niches for long spans of time, feeding on seed ferns such as *Dicroidium*.

XII Old hook snout. The skull of *Hyperodapedon* shows a pair of curved tusks probably used for digging or raking plants to be then pulled into the mouth with a powerful tongue, and sliced with the scissor-like jaws. Note that the lower jaw has been pushed back and should reach the tusks.

(XIII) The Drakensberg Mountains, South Africa. These mountains are built from basalt lavas that erupted 180 million years ago and led to sharp global warming and acid rain. This may be the cause of the early Toarcian extinction event when seabeds were starved of oxygen.

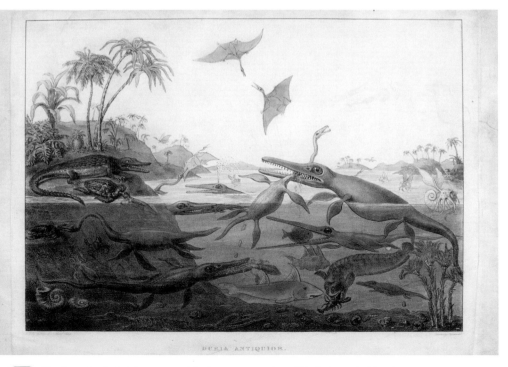

(XIV) 'Duria antiquior'. A famous satirical reconstruction of Early Jurassic life, showing marine reptiles (whale-like ichthyosaurs, long-necked plesiosaurs, crocodiles), fishes, ammonites, belemnites, turtles, and flying pterosaurs above. Henry De la Beche, 1830.

XV Ichthyosaur mother. A famous specimen from the Early Jurassic of Germany, showing the overall dolphin shape and large paddles. Note a baby or two inside the rib cage, and one just making its entrance to the watery outside world at the back of her rib cage

XVI Supreme swimmer. Ichthyosaurs were key marine predators in the Triassic, Jurassic and much of the Cretaceous. They swam by beating their tails and steered with their paddles, and mainly fed by lunging at shoals of fish or squid and grabbing as many as they could.

PART 3

The End-Permian Mass Extinction and Triassic Recovery

252–237 million years ago

CHAPTER 6

The Greatest Crisis of All Time

MUSING ON THE BANKS OF THE MIGHTY SAKMARA

On 12 July 1994, a small party of geologists – including me, Glenn Storrs, a US citizen and then a postdoctoral researcher in Bristol, Misha Surkov, a young lecturer at Saratov State University, our liaison Leonid Shminke, the van driver, and the field trip leader Valentin Tverdokhlebov – reached the high rocky crag of Sambulak. We were in Russia because we wanted to solve a fundamental question about the great crises of the past, and our focus was the end-Permian mass extinction, the largest of them all, when life was nearly annihilated. Fewer than 10% of species survived the devastation, whether on land or in the oceans. What single huge killer, or combination of killing mechanisms, could possibly have brought all of life to the brink?

Early that July morning, we had driven across the roadless steppe to an overlook above the River Sakmara and then walked eastwards along the crest as the land rose towards our destination. As in so many parts of Russia, the steppe seemed endless, stretching for miles in every direction, the metre-high grass billowing gently in waves as the warm winds blew across it. The steppe was not all farmed here, even though there were numerous villages and small kolkhozi (collective farms). We had visited Kolkhoz Pravda (Collective Farm 'Truth') the day before and it appeared a sleepy village with low concrete and wooden cabins, chickens running in the street and a heroic bus shelter with a space rocket on the roof labelled 'Прогрес' (progress), but no sign that buses ever called. Crags such as Sambulak poked up here and there from the otherwise extensive, flat grasslands. As geologists, we interpreted this to mean that the underlying rocks were pretty much flat-lying and that the crags stood proud because they were composed of harder rock that had not weathered away.

We puffed a bit as we climbed, having had a lively party the night before with the local farmers; admittedly, a certain amount of vodka had been drunk. Misha, Leonid and Valentin marched forward resolutely as the ground rose towards the crag of Sambulak, standing proud as a high cliff above the riverbank, while Glenn and I trailed a little behind, somewhat white-faced. Misha explained that we were going up a gentle grassy slope composed of latest Permian mudstones, representing ancient lakes and meandering river deposits. As we panted up to the rocky crag at the top, the slope steepened, and Misha said, 'Now we are in the Triassic.'

We came closer and saw that the crag was 20 m (65 ft) of conglomerate, a rock consisting of pebbles and boulders of older rock that had been broken up, transported and eventually dumped. Such a chaotic accumulation of rocks and boulders indicated an extremely high-energy water flow, something like a flash flood or major hillside torrent. Nothing on such a scale had occurred at all in any of the underlying Permian rocks. All these Permian and Triassic rocks are red in colour and are often collectively called 'red beds'.

As we sat on the ledge at the base of the vertical Triassic crag, Valentin Tverdokhlebov explained that he had worked in this area as part of his PhD work more than twenty years ago. He had noticed the crags all over the region, which extends north and south for several hundred kilometres, and he had understood that these were parts of huge alluvial fans – delta-shaped structures that had built up rapidly down the west side of the Ural Mountains. The Ural Mountain chain extends almost exactly north–south at this point, and it is composed of older rocks, including Silurian and Devonian limestones with fossil corals and brachiopods. Tverdokhlebov had identified the sources of many of the boulders in the Triassic-aged conglomerate and could follow them back upstream into the heart of the Urals.

On the flat plains to the west of the Urals, in his thesis, he had mapped the exact shapes of four or five alluvial fans and found that each fan had a central feeder channel, representing a powerful river hurtling down the steep sides of the ancient Ural Mountains 252 million years ago. Each channel then splayed sideways, forming dozens of smaller channels, carrying many tons of sediment and dumping it as the gradient switched from steep to horizontal on the plain. Some of the mapped alluvial fans were 100 km (60 miles) wide from side to side and extended 150 km (90 miles) west off the side of the Urals. Tverdokhlebov had also mapped the feeder river channels deep into the Ural Mountains where they had begun.

Sitting there under the heavy crag on Sambulak, we debated what might have triggered such violent rainfall, erosion, terrifying water flows and an enormous transport of rocks. And why was it happening at the precise level of the mass extinction?

HOW MUCH WAS KILLED?

Before we went to Russia, I had read everything I could about the end-Permian mass extinction. By 1990, this had been identified as a huge event, much larger than any of the other 'big five' mass extinctions and responsible for the death of 50% of families of marine animals, equivalent to the loss of 90–95% of species. The other mass extinctions – including those at the end of the Ordovician, in the Late Devonian, and even the famous final crisis 66 million years ago when the dinosaurs disappeared – were associated with losses of perhaps 10–20% of families and 50–80% of species. These figures are best estimates, of course, based on our knowledge of the fossil record, which is far from complete. We can count the numbers of families or genera with more confidence than we can count species, and this matters here because it helps to estimate the extent of the end-Permian crisis.

Palaeontologists name their fossils as species, just as biologists name living plants and animals – Homo sapiens, Tyrannosaurus rex and so on. The first bit (Homo, Tyrannosaurus) is the genus name, the second (sapiens, rex) the species name. The names are all Latin and Greek so they can be understood in all lands. Each genus must contain at least one species, and many contain multiple species. For example, the domestic dog is Canis familiaris, the coyote Canis latrans, and the grey wolf Canis lupus. In determining the severity of an extinction event, it is easier to estimate the numbers of families and genera than the numbers of species. This is because there are fewer families than genera, and fewer genera than species, and only one specimen is needed to prove the presence of a family or genus. Knowing the approximate numbers of species in each genus allows palaeontologists to estimate the likelihoods of extinctions at different levels. So, if we know that 50% of families died out during the end-Permian mass extinction, we know this is equivalent, say, to a loss of 75% of genera and perhaps 90–95% of species.

Is this enormous figure of species loss reasonable? It means that only one in ten or one in twenty of all species survived. Work in China has suggested that the global estimate is realistic. In 2012, Haijun Song from the China University of Geosciences in Wuhan collected all the evidence from twenty years of research in the latest Permian and earliest Triassic of South China and identified 537 species of animals living on the seabed. The extinction here occurred in two phases, spaced perhaps 60,000 years apart; the first phase of extinction accounted for 57% of species, the second for 71% of the survivors. When these are added up, 480 of the 537 species died out, and this is 90% of the latest Permian species. All the evidence lines up: 90% of species really did disappear.

But what caused the extinction? What were the catastrophic events that could have driven nearly all of life extinct, not only life on the land surface or in the tropics, but everywhere, from the deepest ocean to the highest mountain and from the Equator to the Poles?

LANDSCAPE CRISIS

That day, back in 1994, as we contemplated the crags around Sambulak, we questioned Valentin Tverdokhlebov about the evidence for the end-Permian landscape crisis in the Russian red beds. He had decided that the sudden event must indicate a renewal of uplift in the Ural Mountains, which had formed through the Carboniferous and Early Permian as two continents collided and fused along a north–south line. Perhaps some deep-seated tectonic movements had caused a final jerk of uplift and so led to a steepening of the mountain slopes and the sudden removal and dumping of millions of tonnes of mountain rocks onto the lateral plains. Why, though, had this mountain uplift episode coincided precisely with the great end-Permian mass extinction? Could there have been a sudden massive increase in rainfall at that point, some freak shift in the climate? He rejected this suggestion, because all the evidence from the rocks showed that climates in Russia at the time were becoming warmer and drier.

Although at that point we couldn't find the answer, we knew that this sharp and short-lived switch in river morphology was significant. Such enormous shifting of rock reflected major landscape change with massive downstream effects. Not only were the mountain rocks rolling and tumbling down the mountain canyons, but water was also rushing past the alluvial fans, carrying millions of tonnes of sand and mud and dumping the sediment all the way downstream to the sea. Tverdokhlebov had identified a process that affected a 300-km (186-mile) front of the Urals; if this occurred on a wider scale, at the level of whole continents or even worldwide, then there would have been a movement of a huge amount of sediment from land to sea. If so, how much of the landscape was stripped of rock and soil, and what were the effects on the coral reefs and other rich ecosystems that occupied shallow waters around the continental margins?

THE COAL GAP

A year later, when we returned to the same area, our sedimentologist Andy Newell, who works for the British Geological Survey, convinced us of what was actually happening. Under his organization, we measured the rock sections at many locations around Sambulak, noting every detail of the rock types and sedimentary structures (such as mud cracks, channels, root traces, burrows and anything else that can provide information about the environment in which the rocks were deposited). We looked at cross-bedding, a common feature in sandstones where during deposition the onward-moving sand is washed or blown down a ripple or dune face. In deserts, the dune face is often high, whereas under a flowing stream, the dunes are usually lower. The orientation of the cross-beds gives the flow direction, and this was the kind of evidence Tverdokhlebov had used to reconstruct the ancient river systems of the earliest Triassic.

Using this detailed process, trotting up and down the badland gorges and covering hundreds of metres of rock sections each day, we were able to settle on what had occurred at the time of river transition: the plants had all been killed by the end-Permian crisis, and without forests and plants in general, the soils were loosened and the verdant landscapes were converted to rocky wastelands by huge erosion. This happens today where forests have been cut down on hillsides – for example, in Bangladesh and Brazil – which is inevitably followed by catastrophic erosion and removal of soil under normal levels of rainfall. The end-Permian crisis reduced the vegetation cover through the extinction of plants, the climates were hot and dry, and the sediments had evidently accumulated in great thicknesses, without evidence for any renewed uplift of the Urals. We published this in 1999 in a carefully argued paper in the rather specialist journal *Sedimentary Geology*.

A year later, palaeontologist Peter Ward from the University of Washington in Seattle and colleagues working in South Africa published a paper in *Science*, a much more prominent journal, in which they pointed out the same switch from gentle meandering rivers to braided streams with alluvial fans and conglomerates at the Permian–Triassic boundary. Braided streams are multiple channels that form and re-form, cross-cutting each other and indicating that a slope is steep with a great deal of sediment being washed down it. They also interpreted this as the result of the loss of vegetation cover and concluded, 'Evidence from correlative nonmarine strata elsewhere in the world containing fluvial Permian–Triassic boundary sections suggests that a catastrophic terrestrial die-off of vegetation was a global event, producing a marked increase in sediment yield.' They had done two things we should have done, which were to review worldwide evidence of what was going on across the Permian–Triassic boundary and to make a stronger argument about the loss of vegetation.

In fact, the vegetation evidence was there. In 1996, a year after our intensive work around Sambulak, Gregory Retallack, an Australian palaeobotanist and sedimentologist who was based at the University of Oregon, and colleagues published a short paper in which they made a profound observation: 'It is a curious fact that no coal seam of Early Triassic age has yet been discovered, and those of Middle Triassic age are rare and thin.' They explained this observation further and named the phenomenon the 'coal gap' – a time when no coal was being formed because there were no trees or forests. There had indeed been abundant trees and forests over much of the world in the Late Permian, and indeed thick accumulations of coal in South China, Australia, India, Russia and even Antarctica, but then nothing for the first ten million years of the Triassic. As Retallack and colleagues noted, their 'coal gap' on land matched the so-called 'reef gap' in the shallow seas, both of them providing evidence for the fundamental destruction of ecosystems by the end-Permian mass extinction.

So, by 2000, there was strong evidence that plant communities on Earth had been largely wiped out by the end-Permian crisis; at the same time, coral reefs, and all their included biodiversity, plus the majority of marine life had also been extirpated. Together with our South African colleagues, we had independently identified a key marker of the nature of the extinction – a huge topographic landscape crisis associated with the death and destruction. It seems that the great conglomerate that ushered in the Triassic wasn't limited to the area Tverdokhlebov had studied in Russia, but also was seen in South Africa, India, Australia and Europe.

Since 1999, geologists have debated the extent of the evidence for this fluvial transition. It may be absent in some parts of the world if there were, for example, regional effects, with perhaps the plant die-off having a bigger impact on landscapes near mountain chains or differing latitudinal effects between the Equator and the Poles. In other cases, there might simply be a gap at the Permian–Triassic boundary where the rock succession is incomplete. The conglomerates are thick and geographically extensive, however, so might be expected to accumulate and still remain to be observed by geologists, although if the erosion was vast, then in some areas of uplands, the conglomerates would have moved off to lower altitudes.

MURDER ON THE ORIENT EXPRESS

At the top of my reading pile before going to Russia was a 1993 book by palaeobiologist Doug Erwin of the Smithsonian Institution, *The Great Paleozoic Crisis*. He had summarized everything that was known at the time and concluded that the mass extinction resulted from what he called the 'Murder on the Orient Express Hypothesis'. In Agatha Christie's famous novel, Hercule Poirot struggles at first to identify who killed the American businessman Samuel Ratchett, but eventually realizes that everyone was involved, each of the twelve passengers separately stabbing him. Erwin argued that the evidence suggested multiple killers at the end of the Permian, each being as responsible as the others for the mass destruction; these culprits included massive volcanic eruptions in Siberia, rising levels of carbon dioxide, a fall in sea level, the loss of oxygen from seabed sediments, global warming and several others. But how could a dozen natural killers coincide in time?

In his book, Erwin connected some of these mechanisms. For example, the volcanic eruptions likely poured huge volumes of carbon dioxide into the atmosphere, and this in turn would inevitably have led to global warming. We are very aware today of the powerful role that carbon dioxide, whether from volcanoes or from our own industrial processes, can have in driving up the air temperature worldwide. Another key fact Erwin noted was that seabeds worldwide had become anoxic at the precise moment of the extinction crisis.

Earlier, in 1992, British palaeontologist and sedimentologist Paul Wignall from the University of Leeds had identified that one of the key phenomena of the end-Permian crisis was that nearly all seabed sediments had lost oxygen – he had seen the effects of this in his fieldwork in Italy and in China and noticed that the switch from oxygen-rich to anoxic sediments was also worldwide. In the stratigraphic record, below the crisis level were pale-coloured latest Permian sediments, full of shells and debris from diverse animals and with rich evidence of burrowing. Above the crisis level, the earliest Triassic sediments were black, devoid of fossil debris and burrows, and sometimes had gold-coloured blocky crystals of iron pyrites (an iron sulphide also known as fool's gold; see page 167). Iron pyrites is formed in anoxic conditions and the black colour of sediments is caused by their rich content of organic carbon. Normally, in the presence of seabed oxygen, diverse forms of life gobble up all the organic carbon and the sediments are pale-coloured; if oxygen is absent, most seabed organisms die and there is nothing to consume the organic carbon.

Could it be that all the killers Erwin had listed could connect together as a single chain of cause and effect?

THE SIBERIAN TRAPS

The connections were becoming obvious for Erwin, Wignall, Retallack and some others. It was well known that there had been volcanic eruptions on a huge scale in Siberia. Great areas of the landscape were dominated by the hugely thick accumulations of basalt, representing prolonged and enormous eruptions of lava from fissure-type volcanoes. These are like the mid-ocean ridge eruptions that continue today and can be seen on land in Iceland. These fissure volcanoes may seem less dramatic than the pointed, or Plinian, volcanoes such as Vesuvius, Etna and Mount St. Helens, but they can produce just as much lava, ash and gases (for more on these volcanoes, see Chapter 10).

In 1990, the Siberian Traps had not been accurately dated, and the best estimates timed the eruptions through much of the Triassic, lasting from 240 to 200 million years ago. Since the 1990s, exact age dating has improved, and the age of the mass extinction and Permian–Triassic boundary was revised to 252 million years ago. Dating of the Siberian Traps eruptions also improved with new sampling and new dating methods; it is now known that the eruptions coincide in age, and with a total span of about a million years, beginning in the latest Permian, before the Permian–Triassic boundary, and ending at some time in the earliest Triassic.

With the dating secure, geologists looked again at the Siberian Traps. Up to 1990, most work had been done by Russian geologists, but after perestroika, many international geologists, including our team, began projects in collaboration with Russian colleagues. Fieldwork in Russia requires careful planning. In Siberia, the localities are under snow and permafrost for much of the year, and in summer people and wildlife are tormented by the giant mosquitoes.

Modern volcanoes are a great source of information about how ancient volcanoes operated. Volcanologists can measure not only the explosive force of eruptions, but also the volumes of lava, ash and volcanic gases that are ejected during an eruption. This is important because this information is scalable, and there are relatively straightforward relationships among the different measures of volcanic output. The Siberian Traps today amount to 4 million cu. km (960,000 cu. miles) of lava, in places making a pile up to 4 km (2½ miles) thick and covering 7 million sq. km (2¾ million sq. miles) of eastern Russia. The volumes of ash and gases produced by the eruptions would have been huge. Although they certainly would have had a worldwide impact, when Wignall did the calculations, he found that, despite their enormous scale, the volumes of carbon dioxide generated by the Siberian Trap eruptions would not have been enough to cause the degree of global warming that happened at the time. Perhaps, he suggested, additional greenhouse gases were released at the same time, such as methane from deep oceanic reservoirs. Another recent suggestion from work in China has been that perhaps there were other volcanoes erupting at the same time in Southeast Asia, contributing to the warming and acid rain, but, if this were the case, how were the eruptions in different locations synchronized?

The gases released during volcanic eruptions include sulphur dioxide, which, when mixed with water in the atmosphere, falls as sulphuric acid. This acid rain kills off forests and acidifies oceans. As we have seen, the widespread loss of trees leads to huge erosion as dead roots are loosened and the soil loses its stability and is washed away by the first rainfall. Acidification of ocean water attacks the calcareous shells and skeletons of molluscs, arthropods and corals.

The already noted extensive anoxia on seabeds was caused by the global warming triggered by the volcanic eruptions. It has been estimated that the temperature rise at this time was as much as 10–15°C (18–27°F), which in the tropics would have raised sea-water temperatures to 34–40°C (93–104°F). Warming of the air feeds into the surface waters of the sea and pushes down the thermocline (the contact of warmer surface and cooler deep waters). If the thermocline goes too low, the usual circulation of cold deep water up to the surface where it captures oxygen and then dives down to the deep seabed elsewhere is stopped. So, warming drove the anoxia, and the combination of warming and anoxia forced marine life, especially in the tropics, either to flee sideways, north or south, or to find a narrow safe zone in the water column, just at the base of the hot surface water, and high enough above the seabed to find some oxygen.

COULD THESE KILLERS REALLY EXTINGUISH NEARLY ALL OF LIFE?

This high-temperature crisis at the end of the Permian is the largest of the hyperthermal events (see Chapter 10), but was it enough to kill so much of marine life? The multiple pressures of heat, oxygen loss and acidification hit some groups of marine creatures hard. Those with calcareous skeletons, and especially those without free-swimming planktonic larvae, might have quickly perished. Fishes, lobsters and sea urchins may have been able to move away from the hot seas, but not all – the Tethys Ocean, for example, was locked by land to north and south. If the heat, acidification and anoxia continued month after month, year after year, this would have taken a heavy toll. Even species that found a safe haven for a while might have been starved out through lack of their usual food or simply been crowded by too many other refugees on the move. In the previous chapter, we noted that the tropics have always been the site of highest biodiversity, and it is evident that a tropical ocean beset by killers, even silent killers like heat, anoxia and acidification, would rapidly empty of life.

And on land? The acid rain killed the forests and plant life in general. As logs, fronds and soil washed downhill and into the oceans, the surviving animals would find less and less food. If the plant-eating insects and reptiles had no food, their predators likewise would have suffered. As in the oceans, tropical belt temperatures reached 34–40°C (93–104°F), and plants and animals would have been on the move. This statement may seem odd for the former, but plants can move in the face of slowly changing conditions. For example, during the ice ages of the last million years, glaciers advanced and retreated many times, so the typical forest trees and other plants of Eurasia and North America marched south and north many times, probably over a few decades or centuries each time. As seeds were thrown down and blown by the wind, an entire forest could move a few metres each year. On land, too, the tropics would have harboured most biodiversity of plants, insects and vertebrates. They would have moved where they could, but if the plants had been largely eliminated, then whole ecosystems would have collapsed and starved. Moving out of the tropics would have led to crowding in the cooler areas and more death.

If the crisis lasted only a few weeks or months, or even a year, life could have bounced back relatively quickly, and the event would probably barely be detected. We see this on a smaller scale after volcanic eruptions today. When Mount St. Helens erupted in 1980, the lava and ash destroyed vegetation and buildings over 600 sq. km (230 sq. miles); witnesses described the apocalyptic levels of destruction and the alien-like grey, devastated landscape. However, after a few years, plants grew back, and thirty years later there is little visible trace of any damage. This short time span might not even register in the rock record.

We know that there were at least two big eruptions that combine as the end-Permian mass extinction, separated by some 60,000 years. In most locations, these two cannot be teased apart, but the evidence from high-quality sites in South China shows two pulses of massive volcanic activity, each associated with sharp temperature rise and acid rain. It is not known, however, how long the warming and acid rain lasted. Perhaps the former lasted longer than the latter, but the destruction of forests by acid rain and coral reefs by ocean acidification were final, and forests and coral reefs did not recover for ten million years. As for the expulsion of life from the tropics, we don't know how long the air and water temperatures remained high, but it might have been for several years, and presumably long enough to carry the extinction levels as high as they were. Only 5–10% of plant and animal species survived on land and in the sea.

Our work in Russia contributed to current models that connect all the climatic crises back to the eruptions of the Siberian Traps. The crisis hit the land first in terms of high temperatures and acid rain, but there are very close connections between land and sea, so as soil and other debris were washed into the oceans, they, too, were being warmed and acidified. The soil clogged the gills and feeding tissues of filter-feeders just as acid attacked their protective skeletons.

There are still a great many questions to be tackled around this most severe of all the mass extinctions, including was it entirely driven by the Siberian Traps eruptions or were other volcanoes involved; where can all the extra carbon dioxide needed to explain the amount of global warming be found; how many killing phases were there; did acid rain strip landscapes everywhere; and are there any reasons to explain why some species at least managed to survive? Whatever the answers, the landscapes of the earliest Triassic were devastated – bleak, hot and devoid of life, with just a few species struggling to survive the harsh conditions and find food. Bad as this was, as we shall see, life did recover.

CHAPTER 7

Triassic Recovery

RECOVERY, EATING MORE
AND THE MODERN WORLD

The biggest mass extinction of all time led to the most profound change in life on Earth. If we look at many of the main groups of plants and animals on the land, we can track their roots back to the Triassic. For example, the oldest frogs, turtles, lizards, crocodiles and mammals all originated at different points through the Triassic. Because we know that dinosaurs have their origins in the Triassic, we can also say that birds, as living dinosaurs, originated then too, so, effectively, all the modern land vertebrates – amphibians, reptiles, birds and mammals – arose in the Triassic. The same is true of some other very obvious parts of the modern terrestrial ecosystem – flies, butterflies and conifers, such as pines, cypresses, firs and monkey puzzle trees, can all be traced back to the Triassic. Just as on land, the origins of many modern groups of marine creatures are in the Triassic – for example, many groups of bivalves, gastropods, crustaceans (lobsters, crabs) and fast-swimming modern-style bony fishes. Why? Was the reason connected with the devastation wrought by the end-Permian mass extinction?

Statistical palaeontologists, such as the great University of Chicago evolutionary theorist Leigh Van Valen, had made this link. In 1982, he described the end-Permian mass extinction as 'resetting' evolution: there was the before and the after, and no other crisis had ever had such a profound effect. At about the same time, Jack Sepkoski, also at the University of Chicago and the founder of big data palaeontology (see page 50), was characterizing the great stages of the history of life in the oceans. He identified a Cambrian phase, a Palaeozoic phase and then the Modern phase, but could not find evidence for a Mesozoic phase, or anything between the Palaeozoic and the Modern. Surprisingly, he found that the Modern phase did not begin after the death of the dinosaurs and ammonites at the end of the later Cretaceous period, but in the Early Triassic.

This was the time when shellfish like mussels, oysters, squid and octopus really flourished, as well as the ancestors of shrimp, crabs and lobsters. Before the Triassic, typical Palaeozoic marine creatures were crunchy and didn't have much flesh: trilobites and crinoids were all skeleton and not much muscle; brachiopods were shell and a few stringy muscles. As American palaeontologist Richard Bambach put it in a 1993 article, seafood improved massively in the Triassic. This marks a huge phase shift, not just a small change here and there, a few new species of this or that, but a major step up in ocean productivity. Animals need food to survive, and somehow the Triassic oceans had a higher food supply than the Permian oceans. On land, too, something similar seems to have happened – many of the new groups, including the dinosaurs and the ancestors of the mammals, were more vigorous and faster-moving than their Permian predecessors. Life on land was more energetic.

So, the question is, what was so profound about the Early Triassic after the devastation that the whole Earth-Life system was reset and speeded up, capturing more of the Sun's energy and consuming more food? To answer this, we must start in South Africa. We head back 252 million years to the moment of the end-Permian mass extinction, and the days after, in the earliest Triassic.

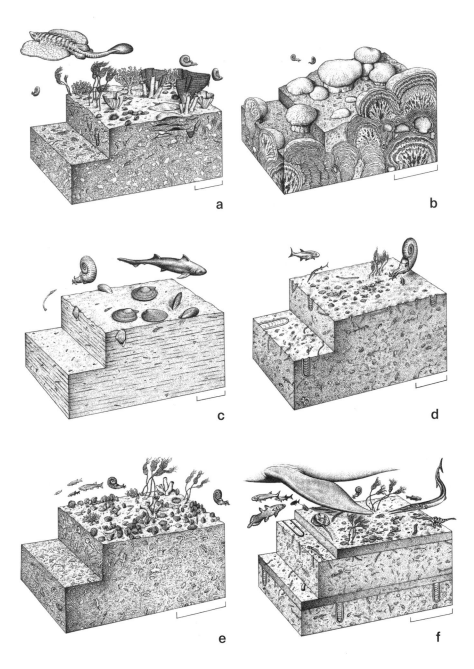

Extinction and recovery. Rich life on the sea floor at the end of the Permian (a) was hit hard, and life recovered slowly at first in the earliest Triassic, as algae dominated (b) or only a few species evolved in black muddy settings (c). Later, more seabed dwellers and swimmers emerged (d–f).

LYSTROSAURUS: JUST SURVIVING

The Sun burns down and the air smells unusual. A pig-like reptile, a female *Lystrosaurus*, emerges from a burrow in the riverbank, having just laid her eggs. Even here, in what is now South Africa, some 9,000 miles (15,000 km) from the centre of the catastrophic Siberian Traps eruptions, the world seems different. The sky is purple even in the middle of the day and it is freezing cold. The *Lystrosaurus* walks along the riverbank looking for plant food. She gathers bunches of ferns and seed fern leaves using her two tusks. She has no other teeth, but her horn-covered sharp jaw edges are highly efficient at slicing the plant stems, operating something like giant shears. She is a messy eater: abundant leaves and stalks fall from either side of her mouth. She makes an effort to tidy up, but not much; some plant-eating cockroaches rustle around on the ground enjoying the feast.

As the days pass, her world changes in unexpected ways. The sky changes colour as if a storm is brewing (see Plate 1). The air temperature shifts rapidly from cold to warm, and then it becomes hotter and hotter. The river dries up and the ground becomes dry and cracked; the bushy plants and scrubby trees go brown. She scents something unusual in the air, a mixture of smoke and acid. In fact, it is becoming difficult to breathe. After a week, food is scarce. Her usual feeding spots are now marked by brown leaves. There have been sudden downpours of rain overnight, usually a welcome event in these dry landscapes, which stimulate plant growth and fill the ponds and rivers. She had emerged early to enjoy the rain, but it stung her eyes. The river is now full and flowing fast, carrying tree trunks and mats of plant debris. On one of the floating mats, another *Lystrosaurus* rushes by, squealing in fear. The water is turbid, full of mud, and soil is being carried away.

We can picture this scene because of pioneering studies of rocks that span the Permian–Triassic boundary in the great Karoo basin of South Africa. Here, palaeontologist Jennifer Botha-Brink and sedimentologist Roger Smith have been working out in detail just what was happening in the rock strata below and above the crisis layer. The rocks here are sandstones and mudstones laid down by ancient rivers and lakes, like the rock successions in Russia. In South Africa, however, the reptile fossil record is more continuous and has been studied since the 1850s.

Our *Lystrosaurus* was witnessing the first days of the crisis when the Siberian Traps eruptions first poured great volumes of sulphur dioxide into the atmosphere, which mixed with water vapour high in the atmosphere to produce sulphuric acid. Sulphuric acid has two effects: first, it forms a vapour that blocks the rays of the Sun and so causes cooling, and second, it falls to the ground as acid rain (for more on sulphur dioxide and its effects, see Chapter 10). In the great London smogs of Victorian times and later, passers-by were nearly asphyxiated by the sulphurous vapours, and I have felt the pricking of acid rain on my eyes during a rare summer downpour in Beijing.

Smith and Botha-Brink identified three phases of killing, two in the latest Permian, and the beginnings of recovery in the Karoo red beds of earliest Triassic age. First, there was a phase of drying associated with loss of plants and small herbivores and carnivores. In the second phase, the main extinction event, there was further drying of the environment associated with more loss of plants, especially the abundant *Glossopteris* tree and horsetails, and the extinction of larger herbivores and carnivores. The third phase, represented by 25–30 m (80–100 ft) of sedimentary rocks above the Permian–Triassic boundary, was a time of extreme aridity and the deposition of wind-blown, desert-like sediment, with the final wave of extinctions and the accumulation of mummified carcasses buried by the sand. Altogether, Smith and Botha-Brink calculate that phases 1 and 2 lasted 61,000 years, based on esti-mated rates of sediment accumulation. This was followed by 50,000 years of no extinctions, then 8,000 years of intense extinc-tion in phase 3.

These levels are followed by rocks identified by Smith and Botha-Brink as representing an early, transient recovery of life. Here, thick conglomerates mark the high rates of erosion caused by loss of trees, which we saw at Sambulak in Russia. *Lystrosaurus* survived through the rigours of drying landscapes and loss of plant food, with two species found in South Africa: *Lystrosaurus murrayi* and *L. declivis*, the first with a short snout and only about 50 cm (20 in.) long; the second with a longer snout and more than 1 m (3 ft) in length. From this level upwards, new amphibians and reptiles appear in the fossil record, such as *Proterosuchus*, a member of one of the new conquering groups of the Triassic, the archosauromorphs, which later included dinosaurs and pterosaurs, and today is represented by crocodiles and birds.

Proterosuchus was a low-slung, vaguely crocodilian-like animal, 1–3.5 m (3–11½ ft) long, with a long, narrow snout armed with back-curved teeth. Its upper jaw strangely bent down and brought its armoury of ten or more front teeth into contact with the shorter lower jaw. It was obviously a predator, and there were plenty of prey *Proterosuchus* could have tackled, including smaller or younger examples of *Lystrosaurus*, as well as fishes and amphibians in rivers and pools. At this point in the earliest Triassic in South Africa, at the time the red mudstones and sandstones of the so-called Palingkloof Member were being laid down, you would have seen signs of a brave new world, the green shoots of recovery.

Within a million years, however, devastation hit again, and *Lystrosaurus*, *Proterosuchus* and many other survivors and newly emerged species died out. Such repeated crises and fitful recoveries spanned at least six to seven million years of the Early Triassic, so it may seem hard to comprehend how such troubled times could have given rise to the modern world. It might be expected that the Earth would bounce back to its normal steady state after an extinction crisis, but there is good evidence for a series of high-temperature crises (hyperthermals; see Chapter 10) throughout the Early Triassic. In a summary of evidence from oxygen and carbon isotopes, geologist Jonathan Payne from Stanford University in California and colleagues showed that temperatures wiggled off course at least four times, with sharp heating episodes at the Permian–Triassic boundary and again one million years later, and then twice again before the Early Triassic was over. Each of these pulses of heating was similar to that at the time of the end-Permian mass extinction, representing a temperature rise of about 10°C (18°F). As life began to recover after each thermal crisis, it was repeatedly cut short by another high-temperature event.

To visit a location where recovery was more or less complete by the Middle Triassic, we go to China and then we can try to fill in the gaps.

THE MIDDLE TRIASSIC IN YUNNAN

My first visit to Luoping was in 2010; it is a bustling county town in the northeast of Yunnan Province, which nestles against the borders of Myanmar, Laos and Vietnam. Having studied the Permian–Triassic boundary in Russia, I was keen to see something of the recovery of life in the first ten million years of the Triassic. The red-bed rock successions in Russia continue upwards into the Triassic, but although there are fossil beds with fishes, amphibians and reptiles, and here and there insects and plants, these are sporadic and not continuous. In South China, on the other hand, there are long sections of limestones, some of them hundreds of metres, even kilometres, thick that document the marine beds through the first ten million years of the period, some deposited in deep water, others on the continental shelf, where life was rich.

I joined a group led by Professors Shixue Hu and Qiyue Zhang of the China Geological Survey based in Chengdu, Sichuan Province, which borders Yunnan to the north, although the remit of the Chengdu Geological Survey Office covers the entire southwest of China, including Tibet. As we drove towards the site, the roads threaded through wide, flat fields of brown earth, many of them planted with tobacco plants. Yunnan has a largely tropical climate and is one of the key tobacco-producing areas in China. Fading posters urged the passers-by to smoke cigarettes as their patriotic duty. In spring, the area is famed for the spectacular display of the bright yellow flowers of rapeseed, also known as canola, the oil-producing relative of cabbage. These spread for miles over the flat farmed land, which is pierced with great limestone hills that are Triassic in age. The limestones here stand high because they are harder than the surrounding rock; the infilling soil is a mix of eroded rock and organic matter built up over thousands of years and enriched by farmers. Every inch of the flat-lying soils between the hills was under cultivation. The industrious farmers had even built small fields by lifting soil into 1-m (3-ft)-wide limestone hollows in which they could grow a single tobacco plant.

We left the main highway and drove up a steep concrete track to the village of Da'aozi. The two-storey houses, all built of concrete and covered with colourful tiles, stood close to the small road. Most had gates into an enclosed yard, so each farming family had a place to store small tractors and farming equipment, and even house a pig or some chickens. The entrances were surrounded by red banners wishing good luck, good health and wealth to all who entered. We walked up a winding path towards the top of one of the hills. There, as I had been told, we could see that half the hill had been removed by geologists. We were met by the custodian of the site who lived in a small house beside the quarry, and he led us up some steps to the first rock bed.

Above us, the excavators had worked into the hillside layer by layer. Their aim had been to take rock samples all the way up through the rock succession, sampling for fossils as they went. Here, there is a 16-m (52-ft) succession of limestones of different kinds through Member II of the Guanling Formation, which is dated as the Middle Triassic at 245–244 million years old, eight million years after the end-Permian mass extinction. Individual limestone beds were about 50–100 cm (20–40 in.) thick and we

Fossil hunting on the large scale. Here, in Luoping County, South China, half a hill has been excavated away to collect fossils from every layer of rock.

could scramble up this natural staircase, one bed at a time. On some surfaces, we saw dozens of small fish specimens, not much larger than garden pond goldfish. On another was an upended box. The custodian removed the rocks holding it down and lifted it aside to show the delicate skeleton of a small ichthyosaur, a dolphin-like predatory marine reptile. At one level there were abundant large burrows of quite elaborate style, forming distinct U-shapes, showing that shrimps were constructing protective homes for themselves in the seabed mud.

LOBSTER LUNCH AT LUOPING

We headed back to the museum in Da'aozi village so that Professors Hu and Zhang could show me all the finds from the quarry above. The 20,000 fossils were all carefully tagged with their bed number. Some were in glass-topped cases and larger ones were attached to the wall, but most were still wrapped in newspaper, piled on shelves and in wooden boxes, ready for preparation and study. We could see that the most common fossils were arthropods such as shrimps and lobsters, as well as fishes and marine reptiles. Among the shellfish were bivalves and gastropods, and rarer forms such as echinoderms (sea urchins and sea lilies), brachiopods, conodonts (fish-like swimmers with small jawbones) and foraminifers (microscopic plankton with hard shells). These are all the typical animals of a shallow sea, and their abundance and the burrows indicate that it was quite shallow water.

Thanks to intensive work by Professor Zhang, some additional groups had been added to the list – belemnites and ammonites (extinct molluscs, typical of the Mesozoic Era, and especially abundant in the Jurassic and Cretaceous). Ammonites occupied coiled shells and swam in surface and deep waters seeking prey that they snatched with their tentacles, while belemnites are bullet-like fossils that are part of the internal calcium carbonate skeleton of a related kind of cephalopod, perhaps similar to a modern cuttlefish, with numerous tentacles for grabbing and horny beaks for snipping up their prey. The Luoping belemnite fossils show arm hooks and even the beaks, evidence of exceptional preservation of soft tissues usually lost in the process of fossilization. Some of the sea urchin specimens even have their surrounding halo of sensory spines preserved. Earlier collectors at the fossil bed had reported plant remains. This was unusual – perhaps the site had been quite close to shore and the plants washed in from the neighbouring land. Zhang also added rare remains of millipedes, more plants and a tooth of a land-living reptile.

Most exciting in the marine assemblage at Luoping were fossils of the top predators. These included abundant examples of the long-snouted fish *Saurichthys*, a heavily scaled snaky fish up to 1 m (3 ft) long, which hunted by the lurk and lunge method. It would remain still, perhaps hiding among water weeds or in the murky bottom waters, then propel itself forward at great speed when a prey animal went by, snapping it up with its long jaws. As we saw

Specimen of an ichthyosaur from the Middle Triassic of Luoping. It has long jaws, and in this specimen, the tail is broken and bent forward over its back.

with *Dunkleosteus* in the Cleveland Shale, such lurking hunters use a marvellous property of living underwater, which is that the act of opening their mouth creates suction and the victim is pulled in (see page 72). Other top predators included marine reptiles such as the ichthyosaur *Mixosaurus*, the nothosaur *Nothosaurus* and the long-necked *Dinocephalosaurus*, all of which fed on fishes and smaller reptiles.

The significance of these fishy and reptilian predators is that they formed a new level at the top of the food chain, something that had not been seen in the Permian; the Luoping biota provides evidence for a new step in the ecological recovery of life after the devastation of the end-Permian mass extinction. Here, eight million years later, the marine ecosystem had rebuilt itself and stabilized, with all the expected levels of life: there were plankton at the base of the food pyramid, which were fed on by molluscs, brachiopods, echinoderms and small fishes. These in turn were fed on by lobsters and larger fishes, which in turn were fed on by the marine reptiles and the predatory *Saurichthys*.

Chinese palaeontologists, most notably Feixiang Wu at the Institute of Vertebrate Paleontology and Paleoanthropology in

Beijing, have made amazing discoveries among these Middle Triassic fishes. Indeed, Wu is something of a fish evangelist and produces outstanding scientifically accurate reconstructions of the new fishes found at Luoping and elsewhere. His drawing of 'saurichthyiform fishes in the Middle Triassic of South China' (see page 125) is deliberately designed to look like an explosion of long, slender fishes with pointed snouts. Some are large, others much smaller. Some have narrow fins, while others have broad pectoral fins at the front or streamer-like fins at the back. Their head shapes all differ in subtle ways.

Overall, this image of the explosion of *Saurichthys* species drives home just how different these marine faunas were when compared to those of the Permian. Life eight million years after the end-Permian crisis was rich, with fleshy molluscs, crustaceans and fishes – Bambach's seafood platter – which was enough to fuel the massive diversification of predatory fishes and reptiles in the warm seas. Wu has discovered all sorts of other wonders in the piscine realm of that time, including *Potanichthys*, the oldest gliding fish. Today, we marvel at the Exocetidae, the flying fishes of the southern oceans, but this mode of life, an adaptation to enable the fishes to escape even larger predators such as reptiles or sharks, had already evolved over 200 million years ago.

The predators feasting on this new wealth of food were not just fast and snappy. Others acquired adaptations for dealing with open tough shells. These durophages (animals that feed on hard food) had pavements of broad, blunt teeth in the roofs of their mouths and could grasp an oyster or hard-shelled crustacean and crunch it flat, then spit out the shell debris and swallow the fleshy body. In the Triassic, reptiles were first to adopt this feeding mode, particularly the placodonts: underwater swimmers with powerful limbs for walking in shallow water and for swimming. Their heads were broad and triangular in shape when viewed from above, providing the extra width to accommodate the crushers and powerful jaw muscles needed for this kind of diet. Although placodonts disappeared at the end of the Triassic, some bony fishes had by then also begun to evolve durophagous habits. Durophagy had existed in Palaeozoic times, but these shell-crushing fishes had reaped much more limited rewards for all their efforts.

If food supply and the overall energy and complexity of ecosystems were increasing in the Triassic oceans, what about on the land? Here, too, there is good evidence for quite a substantial speeding up of life.

CHANGING POSTURE AND THE ARMS RACE

Two major lineages of tetrapods – the archosauromorphs and the synapsids – crossed the crisis boundary from the Permian to the Triassic. We have already met *Lystrosaurus*, a synapsid and a very distant relative of modern mammals, and *Proterosuchus*, an archosauromorph and a similarly distant relative of modern crocodilians and birds. Much more was going on in these first millions of years of the Triassic, although much of it is concealed from us. Fossils are rare and we lack information from many parts of the world and from many age points. When a new group originates, it may do so in one part of the world and examples initially might be quite rare; remarkably, this appears to be true for the origin of the dinosaurs.

Despite a prodigious amount of investigation by dinosaur-mad professors, it has now been discovered that, in fact, dinosaurs had a hidden initial history of as much as 20 million years. The oldest dinosaur skeletons come from the Late Triassic, dated at about 230 million years ago (see Chapter 8), but when we look at the shape of the evolutionary tree, we know that dinosaurs originated at least 250 million years ago. Close relatives – essentially, cousins – of the dinosaur lineage *are* known from the latest Early and Middle Triassic, dating back to 250 million years ago; because these nearest cousins had already differentiated themselves, somewhere the earliest dinosaurs must have existed, small perhaps, rare in their

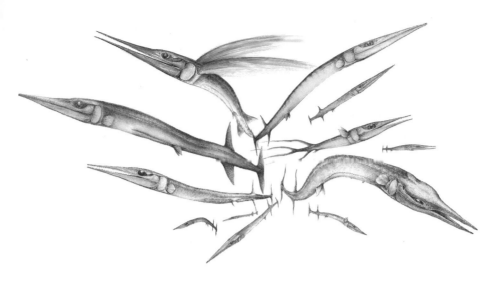

Explosion of diversity. Saurichthyiform fishes from the Middle Triassic of China, showing twelve species of varying sizes and shapes.

ecosystems and maybe restricted to a certain part of the world.

Finding that dinosaurs originated in the Early Triassic massively revises our understanding of the evolution of posture and gait through the Triassic. Some years ago, University of Bristol MSc student Tai Kubo had the idea to look at fossil footprints as evidence of posture. From an animal's trackways, it can be determined whether it walked in an upright manner like modern mammals and birds. Technically, this is called 'parasagittal', meaning the forelimbs and hindlimbs swing back and forwards as they move exactly parallel to the backbone. Early tetrapods were 'sprawlers', meaning the arms and legs stuck out from the sides of the body, and the limb movement during walking was a strange sideways swing of the limbs. The significance of this is that sprawlers can only breathe or run; they can't do both at the same time. This is because as the sprawling animal swings its backbone from side to side, air pumps from the left lung to the right lung but it cannot breathe in and out: modern sprawlers such as salamanders or lizards can sprint for only a short distance before they have to disappear into a crevice or pond for safety.

Parasagittal animals, including humans, horses and ducks, on the other hand, can run and breathe at the same time. Indeed, in fast runners like horses and cheetahs, the back-and-forwards movement of the hindlimbs actually pumps air out and in, which powers sustained fast running.

Kubo and I found that there was a clear switch in the posture of all medium-sized and large tetrapods across the Permian–Triassic boundary. Before the crisis, they were all sprawlers. Afterwards, they were all parasagittal. We speculated that this marked the beginning of an 'arms race', especially because the posture shift happened in synapsids and archosauromorphs at the same time. Although we may be more familiar with the term 'arms race' in a military context, in ecology it refers to both predation and competition. There can be arms races between predators and prey – for example, lions and wildebeest: in the constant tussles between the two, if the lion becomes smarter or faster, the wildebeest must also become faster or smarter to survive. Likewise, competitors for the same resource may experience a similar arms race, whether they are herbivores seeking to maximize their intake of plant food, or predators concentrating on catching as many prey animals as they can – on the plains of Africa, lions eye up the hyaenas, and the hyaenas stay clear, but they are quick to sneak in and deny the lions their feast if they can.

This posture shift and our rather daring suggestion that tetrapods were speeding up would have required higher metabolic rates, meaning they would have had to eat more and breathe more oxygen. Can we really show that these animals were radically changing their whole thermal physiologies at this time?

LIFE IN THE FAST LANE

Metabolic rate in modern animals is usually measured according to their consumption of oxygen, scaled by time and body size (large animals need more oxygen than small animals). We normally

classify modern animals into ectotherms, which have low metabolic rates and acquire heat from the surrounding environment,
and endotherms, which generate some or all of their body heat
from an internal furnace. Reptiles, amphibians, fishes and invertebrates such as molluscs and insects are usually ectotherms;
birds and mammals are endotherms. When and how did endothermy originate?

Some, including me, point to the time of turmoil in the Early
Triassic: the surviving reptiles were moving with a parasagittal
gait, and there is independent evidence that they were endothermic, to some degree at least. Insulation provides the first proof.
Mammals today have hair, birds have feathers – the primary function of these structures is to retain the costly internal heat they
have generated. In 1913, a young British vertebrate palaeontologist,
David Watson (1886–1973), suggested that some of the small car-
nivorous synapsids of the Early and Middle Triassic had hair. The
evidence was that around their snouts there were abundant small
holes through the bone, suggesting rich supplies of nerves. Watson
argued that this showed they not only had muscular lips, as
mammals do today, but also sensory whiskers. Sensory whiskers
are modified hairs, and so from this it may be inferred that they
also had insulating hair all over their bodies. Although this is
debated, some arguing that mammals only became fully endothermic in the Jurassic, the Early Triassic synapsids already had
parasagittal gait, evidence for a diaphragm to help pump their
lungs, enlarged brains and many other features associated with
endothermy.

But what about archosauromorphs? Surely the first feathers
occur only with the oldest bird, Archaeopteryx, in the Late Jurassic,
150 million years ago? Until 1996, this was believed to be the case.
Since then, however, thousands of amazing fossil birds and dinosaurs with feathers have been found in China. At first, palaeontologists accepted that the dinosaurs close to birds had feathers,
but then feathers were identified in other dinosaurs not particularly

closely related to birds. Perhaps all dinosaurs had had feathers from the start and these were simply lost in some giant dinosaurs, which may not have needed them, and in others that had armour plates in their skin? Then, in 2018, a Nanjing–Bristol team led by PhD student Zixiao Yang identified a diversity of feather types in pterosaurs, the distant cousins of dinosaurs. This placed the origin of feathers squarely in the Early Triassic, in line with the synapsids, and was evidence, we argued, that both groups were already endothermic to some extent from that time onwards.

Corroborating evidence comes from bone histology – the study of the cellular-scale anatomy of fossil bone. For example, in a really smart piece of work, palaeontologist Adam Huttenlocker and biologist Colleen Farmer from the University of Utah showed that Triassic synapsids and archosauromorphs had small red blood cells. In modern endotherms, these cells are much smaller than in ectotherms because they are adapted to carry more oxygen to power higher metabolic rates, and the available surface area of blood cells determines how much oxygen can be carried. The minimum capillary dimension matches the red blood cell size; Huttenlocker and Farmer measured capillary size from the fossil bones, which in life carried such tiny blood vessels. If both archosauromorphs and synapsids were endothermic to some extent and had adopted a parasagittal gait during the Early Triassic, this marks one of the most remarkable transitions in vertebrate history, paving the way for the success of birds and mammals in the modern world.

Until recently, the story of the Triassic was seen as relatively simple, with a mass extinction at each end, and the remarkable modernization of life in the sea and on land happening as a single sequence of evolutionary steps in between. However, a mystery perturbation of Earth and life in the Carnian stage, about halfway through the Triassic, is now recognized as a mass extinction event. In the next chapter, we explore how it went undetected for so long.

PART 4

The Late Triassic to Jurassic

237–145 million years ago

CHAPTER 8

The Carnian Pluvial Episode and Diversification of the Dinosaurs

RULE OF THE RHYNCHOSAURS

In the dusty, dry landscape of 230 million years ago, a herd of grey-coloured reptiles marches forward through the dust in what is now Ischigualasto Provincial Park, also known as the Valley of the Moon, in northwestern Argentina. They are rhynchosaurs, distant relatives of crocodiles and dinosaurs, but more pig-like in appearance and measuring about 1.5 m (5 ft) long. This species, *Hyperodapedon sanjuanensis*, has a fat body, round in cross-section like a cow or a pig and evolved to house an enormous, long gut that is needed to digest poorly nutritious plants. It walks on all fours, with stumpy legs, but the digits on each hand and foot are armed with high-sided, narrow claws.

One of the rhynchosaurs stops in a wide and sinuous depression in the ground, evidently a dried-up river. It stands firm and gouges backwards in the soil with one of its hind feet. The claws rip deep into the crust of dry mud and chunks fly backwards; after digging for a few more minutes, there is a great pile of debris thrown up behind it. As it digs, the mud shows signs of damp – the rhynchosaur rotates its body awkwardly, then pushes its snub-nosed head deep, licking the mud for water. Some of the other rhynchosaurs also begin to dig.

One or two on the edge of the dried-up river raise their heads to sniff the hot wind. Their heads are broad across the back, narrowing to a pair of tusks at the front. The overall head shape from above is triangular. From the side, the small eyes, surrounded by folds of wrinkled grey skin, are screwed up to prevent dust blowing in. Some of the rhynchosaurs begin biting fronds from low plants that send deep roots down along the abandoned river to suck water from the rock. The reptiles appear to smile as they chomp: their jaws are curved from back to front, and the lower jaw is a broad blade with worn-down teeth along the crest that cuts up into a groove in the upper jaw, something like a penknife blade folding into its handle.

One of the rhynchosaurs at the edge of the herd raises a squawk of alarm. A slender, two-legged reptile runs round the group. It

The badlands of the Ischigualasto Formation in northwestern Argentina, source of the oldest known dinosaurs, as well as abundant rhynchosaurs.

is about 1 m (3 ft) long and weighs a fraction of the weight of a rhynchosaur. The interloper is Eoraptor, an early dinosaur, and it flashes its array of sharp, curving teeth at the herd, making a chittering sound. Is it showing off, or angry? The Eoraptor knows it cannot tackle any of the adults, and the babies are protected in the middle of the herd. A small, furry mammal scuttles by; it normally hides during the day and comes out at night to hunt insects.

The nearest rhynchosaur emits a deep stomach rumble and lets out a great fart while a copious pile of dung full of undigested twigs and stems drops to the ground. Its diet is poor. Rhynchosaurs feed mainly on a seed fern called Dicroidium, which has a central woody stem and branching fern-like green fronds, and grows around water holes. The reptiles prefer the younger bunches of the fronds, which are up to 1 m (3 ft) tall; it is only after a year or so that the plants transform into 2–4-m (6–13-ft)-high bushes with woody stems, which are less palatable. In either form, the rhynchosaurs have to eat great quantities of Dicroidium in order to survive.

There are conifer trees in the distance, something like modern monkey puzzle trees, with tall, slender trunks and tufts of leaves high above the ground, which are of no interest to the rhynchosaurs because they cannot reach them. A herd of five or six giant Ischigualastia, each weighing about a tonne, barge through the smaller trees, seizing mouthfuls of branches. They are dicynodonts, late-surviving members of a plant-eating lineage that has existed since the Late Permian, with no teeth and a parrot-like beak near the front of their mouths to help rake in plant stems.

This scene in Argentina was common throughout the world at this point in the Late Triassic. Very similar rhynchosaur fossil specimens, species of Hyperodapedon, have been collected in southern Africa, India, North America and Europe. They were highly successful in their day, often dominating their ecosystems, and are represented by hundreds of fossils. It was not to last, however. The dry climates continued through the rest of the Triassic and

Fossil fronds of the seed fern *Dicroidium*, the basic diet of rhynchosaurs and many other herbivores in the Middle and early Late Triassic.

the rhynchosaurs struggled to survive. They had outlived the onset of arid conditions some 232 million years ago, but eventually died out worldwide and were replaced as dominant herbivores in the Late Triassic ecosystems by plant-eating dinosaurs. As we shall see, these dramatic changes in climate were driven by the Carnian Pluvial Episode (CPE), a long-lost mass extinction.

HOW CAN A MASS EXTINCTION BE LOST?

Mass extinctions generally show up in the rocks when there is no doubt that something dramatic has been occurring. For example, a keen fossil collector would note that they find numerous examples of species A, B and C up to a certain level in the rock succession, then after this point these species have entirely disappeared and they find instead species D, E and F. Perhaps, looking further afield, the investigator would identify the same dramatic change from one fossil assemblage to another and make a connection from spot to spot, then eventually link this with evidence from other parts of the world to recognize a global-scale dramatic changeover of species.

There are difficulties with this, of course, especially in moving from the local scale to the global because rock formations may be entirely different between one area and the next. This is most dramatic where the rocks record an ocean setting in one place and a terrestrial setting in another and the fossils are entirely different – molluscs, corals and fishes on the one hand, plants, insects and land vertebrates on the other. Where there are gaps in rock successions, geologists also have to establish whether the sharp faunal change they have identified is in fact real and not the result of a hiatus in deposition representing perhaps a few million years. After such a gap, although the fossils would be different, there might not have been an extinction event but simply evolution at a regular pace with much data loss.

Considerations such as these are what faced an intrepid, and slightly nervous, group of three young researchers back in the 1980s when we each separately spotted something unusual going on during the Carnian stage, the first of three subdivisions of the Late Triassic, dated now as lasting from about 237 to 227 million years ago. I was a fledgling lecturer, in my first serious job at the Queen's University of Belfast in Northern Ireland. Mike Simms and Alastair Ruffell were both finishing their PhDs at the University of Birmingham in the UK.

Simms and Ruffell, at first individually, identified, or even invented, the Carnian Pluvial Episode (CPE). Simms had noted a major extinction among crinoids (sea lilies), a group of echinoderms related to sea urchins and starfish that formed an important part of sea-floor reefs in the Triassic. Meanwhile, Ruffell had observed an unusual sandstone bed near his hometown of Taunton in Somerset, which he had interpreted as evidence for a sharp climatic change. Eventually, in 1987, Simms shared his thoughts about crinoid extinction and Ruffell commented, 'It was raining then. Perhaps the crinoids didn't like the rain.'

Simms and Ruffell continued to discuss their ideas and collect evidence. Ruffell noted that the sandstone layer he had studied in Somerset, called officially the North Curry Sandstone, was very similar to the Arden Sandstone in the West Midlands and the Weston Mouth Sandstone on the Devon coast. At the time, the British Geological Survey was coming to the conclusion that these were all essentially the same rock unit. In comparing them, Ruffell noted that they were all thin, at just a few metres, and occurred in the midst of great thicknesses of red mudstones. The sandstone showed all the signs of having been deposited by water; when Ruffell looked at microscope slides he could see tiny clay mineral grains that on analysis confirmed that the climate had been humid, or pluvial (meaning a time characterized by heavy rainfall).

Simms and Ruffell then leapt across the North Sea to Germany, proposing that this English sandstone was the same as the Schilf-sandstein of southern Germany, which is also a thin sandstone unit in a sequence of mudstones. In all cases, the sandstones contained fossils of fishes and aquatic amphibians and reptiles, and these creatures were very similar, suggesting the same age. In 1989, their paper, 'Synchroneity of climatic change and extinctions in the Late Triassic', announced that 'evidence exists for a significant increase in rainfall during the middle to late Carnian'. The evidence came from Europe and a single locality in Israel, but they boldly claimed it was a global extinction event and highlighted major turnovers, or replacement of species, in the oceans and on land. Had they identified a missing mass extinction? And, if so, how does it relate to the rhynchosaurs of Argentina?

THE HEYDAY OF HYPERODAPEDON

My contribution was to identify a possible mass extinction event in the Carnian, based on comparisons of the fossil records of tetrapods on land and marine ammonites. This began in 1981. In my PhD work, I had chosen to study the rhynchosaur *Hyperodapedon* from the Late Triassic of Elgin in northeast Scotland (see Plates XI and XII). This was essentially the same beast that we met in the Ischigualasto Formation of Argentina and, as previously noted, was distributed almost worldwide as a very successful herbivorous reptile just at the dawn of the dinosaurs. In trying to place *Hyperodapedon* in the wider context, I decided to compare the ecologies of tetrapod ecosystems through the Triassic. I was able to reconstruct the Scottish assemblage of all the other animals living side by side with *Hyperodapedon* and saw that it was important to look at their relative numbers. For example, whereas over the years thirty-five skulls and skeletons of *Hyperodapedon* had been found in various sandstone quarries around Elgin, the other species were each represented by fewer fossils, and sometimes only a single specimen. In fact, I found that out of eighty-seven identifiable specimens of eight different animals from the Elgin Late Triassic, *Hyperodapedon* made up 40% of the fauna. Although these numbers are very provisional because the total sample size was not enough to make accurate claims, at least it allows us to say *Hyperodapedon* was common, while some of the small dinosaur-like beasts with single specimens were not.

As a student, I nervously constructed letters to palaeontologists around the world (these were the days before email). My unusual request to thirty or forty senior colleagues was for specimen counts of their particular Triassic reptile assemblages. This had not been done before – the norm was to collect and describe fossils, and occasionally write a review of the fauna with a list of species names and attributions, but with no indication of the fossil numbers for each species. Amazingly, I received replies from such luminaries as José Bonaparte on the Argentinian faunas, Edwin (Ned) Colbert on the North American faunas, Sankar Chatterjee on the Indian faunas, and James Kitching and Arthur Cruickshank on the South African faunas.

I discovered from the replies that some of the faunas were much larger than those from Elgin – with thousands of catalogued specimens from faunas in South Africa, and hundreds from others. These sample sizes were much more appropriate for statistical analysis, despite the concern that some species, especially larger animals with chunky bones and teeth, might be more often preserved in the rocks and more evident to collectors than the slender bones of tiny animals. On the other hand, fossil collectors often invest special effort in searching out the rarer species, especially the tiny early examples of dinosaurs and mammals, in a bid to find new species or rare specimens. But there is no doubt that the fossil samples I was working with would be incomplete and probably unbalanced as a sample of what actually lived back in the Triassic. I did get some comfort in 2015, when with students we repeated the exercise, checking more recent publications and corresponding (now quickly and by email) with dozens of colleagues worldwide. Their data showed that sample sizes everywhere had increased thanks to thirty years of work, but the percentages of different species in their faunas were more or less the same.

After all the caveats about representative samples, what did the results show? They confirmed that when rhynchosaurs were present in a fauna they tended to dominate – at Ischigualasto in Argentina, they accounted for 39%; Santa Maria in Brazil, 68%; Maleri in India, 57% – all high values, like the 40% figure at Elgin. Admittedly, in North America, rhynchosaurs were rare, and their place was taken by large dicynodonts. Either way, between them, rhynchosaurs and dicynodonts comprised 40–70% of their faunas. This is not unexpected, because they were the key herbivores – the sheep, deer or cattle – of their day. But then they were gone from the fossil record. After a certain point around the end of the Carnian, rhynchosaurs were absent worldwide, and dicynodonts had largely disappeared barring a few late survivors in the United States, Argentina and Poland.

My first thought was that there must have been some serious crisis, especially among the plants that rhynchosaurs and dicynodonts ate. The different reptile faunas were often associated with abundant fossils of the seed fern *Dicroidium*, of which there were many species around the world. It seems that this plant also disappeared at the end of the Carnian, and I suggested that this might explain the demise of the rhynchosaurs. Why did *Dicroidium* disappear? Perhaps there was a climate switch from humid to dry?

I read around further and found that many experts on the marine Carnian rocks of central Europe, especially in the area of the Dolomites in northern Italy and related areas in Switzerland, Austria and southern Germany, had also identified some pretty substantial turnovers among ammonites and other marine creatures in the mid-Carnian.

NUMBER CRUNCHING

In the 1980s, the leading researchers in statistical palaeontology were David Raup (1933–2015) and Jack Sepkoski at the University of Chicago. Sepkoski in particular had made a great compilation of data on marine fossils (see pages 50–52), and he and Raup used these not only to explore both the shape of the history of life through the past 500 million years but also to identify mass extinction events and their magnitudes. For the first time, they were able to quantify these phenomena, such as how many different families or genera of animals were present in the seas at different times through geological time, and the severity of the mass extinctions. Previously, palaeontologists usually only had regional data from their favourite rock successions or general impressions that perhaps one event had been more severe than another.

Raup and Sepkoski, not unexpectedly, identified the mass extinctions at the end of the Permian and at the end of the Triassic, but they also found a peak of extinctions in the Carnian. They argued, however, that this was an artefact or a mistake in counting. In particular, they believed that the end-Triassic mass extinction had perhaps occurred over several millions of years of the Late Triassic, rather than being instantaneous, or that the quality of rock dating was so poor that extinctions were being discovered through a longer span of time than was realistic. However, I felt this Carnian peak of extinctions was real and decided to check the underlying data, focusing on tetrapods, where I had collected the data myself, and on ammonites, which had a rich fossil record; this record had thousands of specimens and could be dated to a much more precise level than Sepkoski had attempted, right down to time units of 2–3 million years each, rather than 5–20 million years. In 1985, I argued in a paper in Nature that the Carnian extinction was real, and that for tetrapods and ammonites it was at least as severe as the end-Triassic mass extinction. At some point in the Carnian, there was a mass extinction event that matched the flourishing and then demise of Hyperodapedon.

Since my work in the 1980s, South American palaeontologists such as Max Langer in Brazil and Martin Ezcurra in Argentina have recognized the so-called Hyperodapedon Assemblage Zone as a world-wide phenomenon, the time when Hyperodapedon flourished. It has been possible to provide exact age dates for this from grains of the mineral zircon in red sandstones of the Ischigualasto and Santa Maria Formations, in the range of 229–228 million years ago, corresponding to the late Carnian. It is therefore likely that other species of Hyperodapedon from India, Scotland and North America are of similar age.

REJECTION ... AND ACCEPTANCE

So, by 1989, Simms, Ruffell and I were feeling quite pleased with ourselves – we had identified a mass extinction event that affected life on land and in the sea; it appeared to be global in extent and as serious a setback for life as the long-accepted end-Triassic mass extinction. Thanks to the careful field and laboratory work by Simms and Ruffell, we even had a climatic model for a major arid-to-humid-to-arid climate shift that could be the driving cause of the crisis. Had we found a sixth mass extinction, equivalent to the established 'big five'? As cheeky young characters, we gave lectures about our great discovery and wrote further papers and reviews to convince people of its existence.

Not everyone agreed, and in 1994 came a serious slap-down. Henk Visscher, a well-established professor at the University of Utrecht in the Netherlands and an expert on fossil plants and ancient climates, wrote a short paper with colleagues as a rebuttal. It had an unusually brutal title: 'Rejection of a Carnian (Late Triassic) "pluvial event" in Europe' (in general, scientists discuss the pros and cons of a case and may couch their rejection in gentler words). He presented detailed analysis of the fossil plants through the Carnian of Germany and agreed that the plants of the Schilf-sandstein confirmed a switch from arid to humid, and then back to arid conditions, as Simms and Ruffell had argued. But Visscher believed this was just a local habitat switch within an overall dry time. In particular, he noted that independent dating evidence showed that the Schilfsandstein was diachronous, meaning it was of different ages in different places. If so, then it would be only an indication of local occurrences of wet ground and rivers and nothing more. As Simms later ruefully noted, 'with our idea seemingly destined for obscurity, we each moved on to other things'.

But in 2016, Simms, Ruffell and Paul Wignall, now at the University of Leeds, looked again at the spread of data. Thanks to the work of geologists in many parts of the world, they were able to identify evidence for the CPE in North America, Japan, the Middle East and Europe. This was indeed a worldwide event, as Simms and Ruffell had argued in 1989. Earlier, in 2015, young Italian geochemist Jacopo dal Corso had identified the driving factor of the CPE, namely huge volcanic eruptions on the Wrangellia land mass, which now forms part of the west coast of British Columbia in Canada. Dal Corso showed that these thick basalts were of Carnian age and were erupting just as the sandstones indicating humid conditions began to be laid down. He also analysed the geochemistry of many rock sections and showed the same oxygen and carbon isotope signals that mark other extinction events, such as the end-Permian event. We encountered the oxygen isotope palaeothermometer when discussing the Late Ordovician mass extinction (see Chapter 3). Sharp changes in oxygen isotopes mark sudden shifts in temperature, and these are often matched by rapid changes in carbon isotopes, too. These changes document how much light carbon is locked up in living plants and animals and how much is in the atmosphere, and so can record the consequences of large extinction events, or when productivity is high – times when there is plenty of plant food on land and abundant plankton in the oceans.

The new work showed that the CPE lasted for about a million years, from 233 to 232 million years ago, and that there were four or five isotopic anomalies, meaning times of increased volcanic eruption that led to sharp warming. As the Wrangellia volcanoes erupted, carbon dioxide and other greenhouse gases drove rapidly rising temperatures, just as during the end-Permian mass extinction (see Chapter 6). As then, warming and acid rain affected life on land as well as in the oceans. An added effect at this time was enhanced rainfall, which was also driven by the sharply rising temperatures. As temperatures rose, water evaporated from the land and ocean surface, forming great vapour clouds that swept along the coasts causing torrential, tropical-type rain storms. With additional rainfall, in coastal areas of the supercontinent Pangaea at least, rivers flowed, depositing sandstones and encouraging the growth of damp-loving plants.

In 2020, Jacopo dal Corso led a large group of authors, including Ruffell, Wignall and me, as well as many younger researchers, in the writing of a review paper in *Science Advances* entitled 'Extinction and dawn of the modern world in the Carnian (Late Triassic)'. Although hidden for thirty years, the CPE had finally emerged as a new, and previously undiscovered, mass extinction. Does it measure up with the 'big five'? Probably not. According to our calculations of extinction rate, the end-Triassic mass extinction accounted for losses of 45% of marine genera, and the end-Cretaceous crisis about 50%. The CPE was responsible for 35% of losses, which is more in line with a class of large extinction events rather than mass extinctions.

THE DAWN OF THE MODERN WORLD

The title of the 2020 paper referred to the 'dawn of the modern world'. We made this claim because some key elements of the modern world can be tracked to the aftermath of the CPE. We had identified this as the time when dinosaurs finally came into their own – they had originated in the Early Triassic (see page 124), but remained at very low diversity and small body size, only becoming ecologically important in the late Carnian and after that time. It may seem unexpected to treat dinosaurs as part of this modern world, but living around these burgeoning early dinosaurs were the first turtles, lizards, crocodiles and mammals, as well as modern groups of insects including flies, beetles and butterflies, and the first modern-type conifers. And as birds are living dinosaurs we could say these early dinosaurs are also part of the modern-day fauna.

There was a large burst of innovation at this point among insects, with many new plant-eating modes emerging for the first time, from leaf chewing to sap sucking. In the oceans, too, the end of the CPE coincided with the origins of modern types of plankton, such as dinoflagellates with their calcareous (lime-rich) shells, and the beginning of a major switch in the carbon cycling system in the oceans. Up to the CPE, carbon cycling was limited mostly to the continental shelf – carbon was captured in the calcareous skeletons of corals, brachiopods, molluscs and arthropods; when they died their skeletons fell to the seabed from which the constituent carbon and calcium could be recycled. This shelf-limited carbon factory was massively disrupted when sea levels went up or down – for example, when sea levels fell, much of the continental shelf became land and the carbon factory was hugely reduced. As sea levels rose, it increased. From the Late Triassic onwards, the carbon factory stabilized in its modern form because most of the carbon came from the calcareous skeletons of the new plankton organisms and so fell to the seabed not only on the continental shelves but also in the deep oceans. Changing sea levels do not disrupt this system – and it all began after the CPE.

We noted other major perturbations in the oceans. Ammonites show a turnover, with old species disappearing and new species emerging, just as we had known back in the 1980s. Also, as Simms had pointed out, there was a major extinction of crinoids. Perhaps more importantly for the modern world, this was when modern-type coral reefs first diversified. Until then, most reefs in the earlier Triassic had been dominated by microbes, crinoids and other reef-builders.

Much of the modern world, then, traces back to this unusual time in the Carnian when the climate flip-flopped from dry to humid and then back to dry again. The CPE marked a major reset of life and even though the climate change on land was to more arid conditions, nonetheless this stimulated a remarkable flourishing of modern-style plants, insects and vertebrates. In general, though, the Triassic was a time of turmoil, and there was more disaster to come just as these new ecosystems, dominated by dinosaurs, were becoming established.

CHAPTER 9

The End-Triassic Mass Extinction

THREE MEN IN A BOAT

The year 1819 was a big one for the Reverend William Buckland (1784–1856). He had just been appointed to the Chair of Geology in Oxford and was perhaps the first to see the geological evidence of the end-Triassic mass extinction. Long after his time, this event was identified as one of the 'big five' mass extinctions and was implicated in the origin of the dinosaurs, or at least identified as the trigger that set them on their course to ecological dominance. But the event was a mystery for many decades after 1819 – what were its effects? Was it a single sharp event, multiple smaller events or even a long-drawn-out process? And what caused it: was it an impact, or massive volcanic eruptions? As we shall see, the various pieces of evidence came together only in more recent decades.

But, back to William Buckland, who was one of life's eccentrics. His most famous achievement was to have eaten his way through most of the animal kingdom; he stated later in life that he enjoyed most of the meats he had tasted, but despite all the best efforts of his cook, she could not make bluebottle or mole palatable. Born in Axminster in Devon, he later befriended the famous Dorset fossil collector Mary Anning (1799–1847) from whom he bought fossils. He trained as a theologian in Oxford, but attended lectures in mineralogy and chemistry, and was appointed Reader in Mineralogy at Oxford in 1813. He ascended to the new Chair in Geology in 1818 and became a Canon at Christ Church Cathedral, which entitled him to apartments in the college (this mixing of church and scientific duties at a leading university was common at the time and he later became Dean of Westminster, a very senior post in the Church of England). He was an avid geologist and palaeontologist, travelling widely across England and indeed more extensively in Europe in search of interesting fossils. In the early 1820s, he described Pleistocene bone caves full of Ice Age hyaenas, cave bears and cattle, and also named the first dinosaur ever to be named, *Megalosaurus*, based on bones he had acquired over the years from quarries in the Jurassic rocks around Oxford.

In 1819, Buckland visited the Aust Cliff locality on the banks of the Severn Estuary, near Bristol, which provides a classic rock section spanning the Triassic–Jurassic boundary; the particular rocks Buckland examined, and which are seen more widely across Europe, eventually proved to be crucial in understanding the end-Triassic mass extinction event. One of Buckland's companions was his friend William Conybeare (1787–1857), then Rector of Sully in South Wales and earlier a student at Christ Church, Oxford, which is where they had met. They were accompanied by an aspiring young geologist, Henry De la Beche (1796–1855), then in his early twenties and later to be the first director of the Geological Survey of Great Britain.

Their detailed descriptions of the locality were published in a splendid monograph in 1824, but the paper had been first read to the members of the Geological Society of London in a series of lectures at their evening meetings from December 1819 to March 1820. De la Beche was an accomplished artist and drew the views and sections in the monograph. The view of Aust Cliff in the paper shows the entire length of the 3-km (2-mile)-wide cliff, coloured deep red in the lower part, and grey and blue-black towards the top. One close-up section depicts gypsum deposits near the base of the cliff being carefully examined by a delightful 1820s gentleman in black coat and top hat; at some point all three, or at least De la Beche, must have ventured out onto the waters of the Severn Estuary in a boat to capture the long view. I like to think of them, perhaps like a scene from Jerome K. Jerome's *Three Men in a Boat*, having uproarious adventures as they took notes, made sketches and argued and bickered about the geology they saw, and who should master the oars, or who was most likely to capsize the boat.

THE GEOLOGY OF AUST CLIFF

The key feature at Aust Cliff is the sharp line between the red mudstones at the base, and the black and grey mudstones and limestones above. What Buckland and his companions reported in their 1824 memoir is remarkably prescient when viewed with modern eyes, especially because geology as a science barely existed in the 1820s and there was no comparative framework in which to place their observations. They identified the red-coloured rocks as 'New Red Sandstone' and the grey and black rocks as 'Lias'. We now call these Triassic and Jurassic respectively (the word 'Triassic' was not coined until 1834, ten years after they published). The red Triassic sediments were formed from red dust blown by the wind over lowland areas, which had accumulated over millions of years to form great thicknesses of mudstone.

In the red rocks at the base of the cliff, Buckland had spotted gypsum, a useful mineral containing calcium and sulphur, widely used then and now to provide an intense white colouring to paper and toothpaste. Today, gypsum accumulates in sabkhas, which are brackish-water bodies seen in hot climates, such as along the north coast of Africa or in the Middle East. Evaporation rates under the burning Sun are high and seawater is drawn underground

The famous geological section at Aust Cliff, near Bristol, southwest England, showing Triassic rocks (red to black) below and the Jurassic (grey-coloured) in the undergrowth at the top.

into the evaporating basin, where great deposits of salt accumulate over time. The evaporation concentrates the brines soaking the sediment, causing the salt to precipitate out and grow into lumpy, nodular masses. The presence of gypsum confirms very hot temperatures, as Buckland and Conybeare would have known, and we now understand by the fact that Great Britain lay just north of the Equator at the time.

Buckland and colleagues identified all the grey, black and blue-grey bedded rocks at the top of Aust Cliff as marine sediments, partly because of the nature of the rocks, but especially due to their fossil content, which included abundant bivalves (particularly oysters), gastropods, ammonites, crinoids, echinoids and bits of crabs and lobsters. All of these fossils were easy for them to compare with modern examples, the majority of which are marine species. But what about that sharp contact between the red Triassic and black and grey Lias rocks?

THE RHAETIAN TRANSGRESSION

The sharp colour change in the rock layers marks a shift from terrestrial to marine conditions. Buckland and Conybeare knew the rocks were ancient, but they did not have access to exact ages. They also knew they were full of fossils, but did not know that this represented a time of mass extinction. However, they did recognize that they were looking at an amazing and sharp environmental shift, what we now call a marine transgression – a time when the sea floods over the land.

It later became evident that this sharp environmental switch occurred not only in southwest England, but also throughout much of Europe, across Germany to Poland, and south and west into Switzerland and France. This was clearly an important event in Earth history and is grandly titled the Rhaetian Transgression, when Europe, at least, turned from land to shallow sea, one of a much larger set of geological events that triggered the end-Triassic mass extinction. The word 'Rhaetian' is an odd term; it comes from the German Rhät (or Rhaet) named after the Rhätisch Alpen on the borders of Switzerland, Austria and Italy, which in turn were named after the Roman Raetia Province. The Rhaetian was named as a formal time interval in 1861 and is the international term used for the final stratigraphic stage of the Triassic, just below the Triassic–Jurassic boundary, corresponding to the mass extinction, 201 million years ago.

The Rhaetian Transgression changed the whole landscape, sweeping in from the open ocean to south and west. As the seawaters swirled about, they gouged and eroded the red beds below, tearing up soft balls of the sediment and rolling them along. Occasional storms blew up under the tropical skies, sweeping inshore and piling up the seawater around the remaining land, and then the waters swept back offshore under gravity, picking up debris from the shore and from the shallow seas. Hunks of mud, shells, bones and teeth were entrained by the roiling waters and carried tens or hundreds of metres downslope until the rate of flow slowed and the debris fell to the sea floor.

When we take our students to Aust Cliff, we tell them the story of the amazing span of time encapsulated at this remarkable location. We now know that the Triassic red beds with gypsum were laid down some 220 million years ago, and the overlying Lower Jurassic rocks 195 million years ago, so they are looking precisely at the Triassic–Jurassic boundary, the point at which one of the 'big five' mass extinctions occurred. This was a time of immense tectonic activity in the Earth's crust and the opening of the North Atlantic, when volcanoes erupted nearby, showering debris, changing climates and killing life. The students wander off, unimpressed, looking for fossils of shark and ichthyosaur teeth.

BONE BEDS AND COPROLITES

These were early days in the history of geology, and pioneers like Buckland and Conybeare were often the first to do things: in their 1824 paper, they introduced the term 'bone bed', using it, as we do today, to mean a layer of rock that is rich in fossilized animal bones, teeth and scales (in their case, fish and reptile remains). The important point about the Rhaetian bone bed is that it lay at the base of the marine rock succession. Buckland and Conybeare noted that 'Mr. Miller of Bristol has in his collection from the bone-bed at Aust Passage many large tuberculated bodies, extremely compact, and of a jet-black colour, which were probably connected with the palates of some very large cartilaginous fishes.' The bones in Mr Miller's collection were black, shiny, rounded specimens about the size of a gingernut biscuit and later recognized as the grinding tooth plates of the lungfish *Ceratodus*. Their colour indicates intense swamping of the bones with the element phosphorus (phosphatization), later to be recognized as a key indicator of the catastrophic geochemical changes in the oceans at the time of extinction (see page 65). Buckland and Conybeare noted the bone beds at other locations along the Severn, characterizing these as 'loaded with the scales, teeth, palates and bones of fishes, and the bones of many gigantic reptiles ... they abound in iron-pyrites, and in rolled fragments of bone and of differently coloured clay-stone, of which the darker varieties resemble the bone in colour'.

The Rhaetian bone bed at the bottom of the marine succession is often directly in contact with the underlying red rocks. In some locations, we have found crayfish burrows cut into the top of the red mudstones, formed after the Rhaetian Transgression and filled with bones and teeth. These lobstery small creatures were living on the seabed, burrowing for safety into the mud and hunkering down as the storm surge ebb currents rushed by. The current dumped its bedload of transported teeth and bones on top, then the crayfish scrabbled up through the debris, packing the bone bed fragments behind them into their burrows.

Buckland later described phosphatized bones and fossil excrement from the Lias of Dorset, largely thanks to the remarkable fossil-hunting efforts of his friend, Mary Anning. In 1829, he wrote a thorough account, describing their shapes, sizes, internal compositions and possible producers, and inventing the new name 'coprolites' for fossil faeces. Despite being a man of the cloth, Buckland relished discussion of the digestive habits and productions of ancient life and wrote several papers about coprolites. In 1830, to tease his older friend, De la Beche made a delightful watercolour of the life of Dorset in Early Jurassic times called 'Duria Antiquior', illustrating ichthyosaurs, plesiosaurs, sharks and bony fishes gleefully eating each other, and snapping at ammonites and belemnites. Leathery winged pterosaurs fly above. What makes the image (see Plate XIV) perhaps less idyllic, is that all these creatures are defecating copiously, producing the coprolites that so delighted him.

THE EXTENT AND POSSIBLE CAUSES OF THE MASS EXTINCTION

Buckland and Conybeare saw the local effects of the end-Triassic crisis, but later work worldwide has cemented a clearer picture. In the oceans, about 60% of species disappeared, with the losses widely distributed. Among ammonites, the major Triassic group, the ceratitids, declined through the Late Triassic and disappeared. There was a massive reef collapse, explained by substantial ocean acidification, with the disappearance of many species; reefs did not recover for three to four million years into the Jurassic.

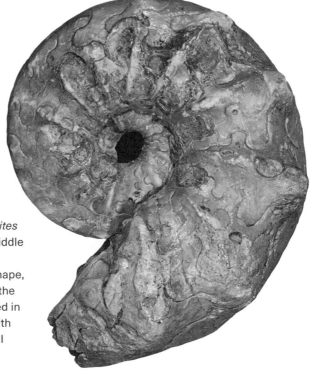

The ammonite *Ceratites nodosus* from the Middle Triassic of Germany, showing its coiled shape, ribs, and remains of the shell. The animal lived in the final chamber, with tentacles, like a small octopus.

Among marine vertebrates, the conodonts (see page 121) finally disappeared. Perhaps more significant ecologically was the collapse in the diversity and variety of marine reptiles. As we saw in Chapter 7, marine reptiles such as ichthyosaurs, nothosaurs and placodonts were widespread and dominated marine ecosystems, but they went into decline through the later Triassic. Only a few ichthyosaurs and placodonts seem to have survived until the Rhaetian – bones of ichthyosaurs, including some giants, as well as placodont teeth, have been found in the basal Rhaetian bone beds. This was a huge turnover among marine reptiles across the Triassic–Jurassic transition, a major evolutionary bottleneck, a time when diversity was pinched off. Ichthyosaurs and plesiosaurs survived and then flourished in the Early Jurassic, although they represented a much-reduced number of founder lineages, as shown by University of Bristol MSc student Philippa Thorne and colleagues in 2011.

On land, there was a substantial turnover among plants, with many common Triassic species disappearing. The impact on insects is uncertain because of limited fossil evidence, but among the reptiles, this was a hugely significant time. As we saw in Chapter 8, the dinosaurs had substantially expanded in diversity following the Carnian Pluvial Episode, but the key predators in the Late Triassic were rauisuchians, large flesh-eating archosaurs related to modern crocodilians. These and other groups such as the fish-eating phytosaurs and root-grubbing aetosaurs also disappeared, either at the very end of the Triassic or at some point during the Rhaetian. The demise of these archosaurs seems to have enabled new groups of dinosaurs to diversify in the Early Jurassic, including larger predatory dinosaurs and armoured dinosaurs, some like the ankylosaurs, with bony armour plates set in their skin, and the stegosaurs, with upstanding plates and spikes down their backs and along their tails.

For many years, palaeontologists have debated the suddenness of the end-Triassic event, and this has been tied to debates about the causes, whether gradual environmental change, asteroid impact or volcanic eruptions. One of the first detailed investigations was by Edwin (Ned) Colbert (1905–2001), a renowned expert on Triassic reptiles who excavated worldwide and wrote many popular science books. Colbert argued that there were slow environmental changes through the transition from the Triassic to the Jurassic and a general reduction in the diversity of habitats. Some of the physical global changes, including the flooding of what had once been land by shallow seas (the Rhaetian Transgression) and the overall reduction in habitat variety, Colbert suggested, were enough to explain the extinctions.

Others, however, believed that this was not enough, and later suggestions were more dramatic. Ever since the 1980s, when asteroid fever was at full pitch (see page 193), attempts were made to explain the end-Triassic mass extinction by extraterrestrial impact. The Manicouagan crater in Quebec was the right size and the right age ... or so it was thought. The crater, 100 km (60 miles) across and about the size of the end-Cretaceous crater (see pages 196–97), had been punched into the Earth's surface and could be seen clearly in satellite photographs. However, later studies showed that the Manicouagan crater had been formed 13 million years before the end of the Triassic, so the asteroid impact that produced it could not be the cause.

THE CAMP ERUPTIONS

After 1999, the focus turned to massive volcanic eruptions. Before then, volcanic causes had not been seriously considered for the mass extinction because the smoking gun was not recognized, even though it was right under the noses of geologists and might well have been expected. It had long been understood that through the latest Triassic, the North Atlantic Ocean had begun to unzip. Europe and Africa on one side, and North America on the other, began to move apart at the usual continental drift rate of about 2.5 cm (1 in.) per year. We can see what this must have looked like in today's Great Rift Valley, which slashes down the eastern side of Africa from Ethiopia in the north to Mozambique in the south. Along this 7,000-km (4,300-mile) cut through the Earth's crust, there is constant volcanic activity as new crust bubbles up, noxious gases are vented and long lakes, sometimes highly saline, fill the gaps. In the Late Triassic, this is what lay a few hundred miles west of Bristol in the UK. The unzipping Atlantic was at first not filled with seawater, but long faults, or cracks, ripped open and magma bubbled up. Elongate freshwater rift lakes formed on the North American side, and these can be seen parallel to the east coast from Nova Scotia to Virginia. On the European side, the North Atlantic rifting propagated faults all across the Bristol area and may have triggered the Rhaetian Transgression by generally lowering land levels so that the sea flooded across.

The volcanic driving force of the continental split can be seen in the Central Atlantic Magmatic Province (CAMP), a huge zone of lava measuring about 11 million sq. km (4.2 million sq. miles) and extending on one side along the eastern coastal strip of North America, the Caribbean and central Brazil, and on the other, the UK, Iberia and western Africa. Great slabs of basalt lava can be seen in Morocco and down the eastern North American coast, especially in the Palisades, near New York City. The volume of lava is estimated to have been about half that of the Siberian Traps eruptions that drove the end-Permian mass extinctions (see Chapter 6), namely 2–3 million cu. km (480,000–720,000 cu. miles) compared to 4 million cu. km (1 million cu. miles).

Why was this huge emplacement of lava not recognized and the link made to the late-Triassic mass extinction until 1999? After all, this was one of the most devastating times of climate change in Earth history, in the midst of the age of the dinosaurs, and the changes were clearly evident in North America and Europe. There are two reasons for this. First, the dating had not been precise enough to be sure whether the eruptions coincided with the extinction or not. In 1988, American geoscientists Michael Rampino and Richard

Stothers compared all the age dates for the eastern American basalts and linked them to the end-Triassic mass extinction. Then, in 1999, geologist Andrea Marzoli from the University of Geneva and colleagues made the geographic leap from eastern North America to Morocco and Brazil, naming the Central Atlantic Magmatic Province and making a strong case for one unified zone. They pointed out that the lava types in all these areas were chemically the same, they all indicated similar orientations of magnetization, and most importantly, shared the same age, of about 200 million years ago.

The link between CAMP and the end-Triassic mass extinction was made crystal clear in 2007 by geochemist Jessica Whiteside from Columbia University, New York and colleagues with their study of the intimate links between the volcanic lava flows in North America and massive turnovers in plants and vertebrates at the same time. Like the Siberian Traps at the Permian–Triassic boundary, the huge CAMP eruptions belched carbon dioxide, methane and water vapour into the atmosphere, generating both an intense global warming episode and acid rain at the same time.

The CAMP eruptions continued for about 600,000 years, beginning before the end of the Triassic, peaking at the boundary and continuing briefly into the Early Jurassic. During this time, there were four pulses of high rates of eruption, separated by more quiet periods, as shown by pioneering uranium-lead dating work by geologist Terry Blackburn, then at Massachusetts Institute of Technology, and colleagues in 2013.

ONE EVENT OR SEVERAL?

Are we looking at a single, focused burst of extinctions worldwide just at the end of Triassic time or were there several pulses of species loss? Further, how do these crises correspond to the Rhaetian? The debate has been complicated by the fact that the Rhaetian Transgression was restricted to Europe, but the extinctions were worldwide. There is also a serious question about the duration of the Rhaetian time stage – was it five to eight million years long, or little more than a geological blink of an eye?

In 2012, geologist Guillaume Suan from the University of Lausanne in Switzerland and colleagues made a strong link between the CAMP volcanic activity, extinctions and the Rhaetian bone beds. They showed that the beds were not simply chance accumulations of bones that resulted from physical processes such as storms, storm surge ebb currents and piling up under gravity and waning current velocity. There were chemical causes, too, most notably an episode of low oxygen in the oceans (anoxia) associated with shifts in the carbon isotope signal (see page 141) and high levels of phosphorus derived from buried organic matter.

Bones and teeth are composed of the mineral apatite (calcium phosphate), and as we have seen, everything in the Rhaetian bone beds was black and shiny, meaning bones, teeth, coprolites and even sometimes random lumps of sediment were heavily infused with phosphate. This kind of intense phosphatization can only happen in the presence of oxygen, whereas the bone beds now lie in black mudstones, indicating anoxic conditions. Therefore, the intense accumulation of phosphorus occurred before the bones, teeth, coprolites and lumps of sediment were transported offshore and downslope into deeper waters. Then, as now at warmer latitudes, the seawaters around the coasts were layered, with oxygen-rich waters at the top and oxygen-poor waters below, with little mixing. The division between these water layers is a thermocline, a transition layer in the water column marking a sharp temperature change (see also page 109).

Suan and colleagues noted that there were several isotope spikes through the British Rhaetian. Here, we have to introduce some stratigraphic terminology. The Rhaetian in the UK comprises two formations: the Westbury Formation below and the Lilstock Formation above. The Lilstock Formation is further subdivided into the Cotham and Langport Members. The isotope spikes are at the base of the Westbury Formation; at the contact of the

Westbury Formation and the Cotham Member; and at the Triassic–Jurassic boundary. Suan and colleagues suggest all three shifts in carbon and oxygen isotopes mark pulses of eruptions of the CAMP lavas, each of which pumped huge amounts of carbon dioxide into the atmosphere, indicated in the carbon isotope shift, and pushing the temperature up by 5°C (9°F) or so, indicated by the oxygen isotope shift.

The reality of this first extinction event at the base of the Rhaetian has been debated. However, geologist Manuel Rigo from the University of Padua in Italy and colleagues argued in 2020 that a comparison of rock sections worldwide showed that a substantial extinction had occurred. They noted geochemical indicators of crisis, including sharp spikes in the oxygen and carbon isotope signals at or very close to the base of the Rhaetian, and this was associated with extinctions. Most striking losses were among the ammonoids, and this was the time when their diversity plummeted most severely, with the loss of all the dominant ceratitid ammonites, followed by low biodiversity until the Jurassic. Bivalves also showed major declines at this time, much greater at the end of the Norian (the age before the Rhaetian) than they experienced at the end of the Triassic. Sharp extinctions were also seen among the radiolarians, a group of plankton, as well as diversity drops among conodonts, reef corals, fishes and marine reptiles.

Also in 2020, importantly, University of Leeds palaeontologists Paul Wignall and Jed Atkinson were able to use the excellent British Rhaetian rock sections to clarify what went on in the second and third pulses of extinction. They showed that the second event, at the transition from the Westbury Formation to the Cotham Member, was marked by extinctions of many bivalves and ostracods (small aquatic shrimp-like creatures protected by a paired shell), and the third, at the Triassic–Jurassic boundary, by the loss of further bivalves and ostracods and the last gasp of the conodonts. These extinction pulses were matched by major changes in the plant records on land.

In Britain and much of Europe, thanks to the Rhaetian Transgression, we have excellent records of shallow marine seas through this time, full of fossils and highlighting events in detail. Little did Buckland and Conybeare realize the global importance of their observations at Aust Cliff back in 1819. It's no wonder that the three events now discerned in the British and European Rhaetian might have been fudged because in many places the rock record is incomplete or poorly dated, and it's still debated just how far apart these three episodes are in time.

Current overviews of the best available evidence on rock dating estimate that the Rhaetian Stage is eight, five or even less than one million years in duration. Various estimates of duration based on radiometric dating suggest that the second two extinction peaks might be much less than half a million years apart, but what we don't know is whether the Rhaetian as a whole represents a total of one million years or five million. The standard date for the beginning of the Rhaetian is 205.7 million years ago, but one suggestion in 2020 was that the start date is 201.7, making the whole Rhaetian succession that Buckland saw in Aust Cliff a mere 200,000 years, with the isotope spikes possibly explained by three pulses of CAMP volcanism.

Whatever the debates around dating, the end-Triassic event marked a major turnover among ammonites and marine reptiles in the oceans, as well as among dinosaurs and other reptiles on land. As we look in the next chapter at how life bounced back in the Jurassic, we consider in more detail the 'common model' of hyperthermal crises that has emerged in recent years for many of the mass extinctions through the Permian and Triassic. These events, and many others, were driven by volcanic eruptions, and especially by global warming and acid rain. This model matters because we are living today through one such event.

CHAPTER 10

The Universal Hyperthermal Crisis Model

RECOGNITION OF THE UNIVERSALITY OF HYPERTHERMALS

The word 'hyperthermal' is relatively new. Meaning 'unusually high temperature', it is used in the context of environmental temperatures and by ecologists and geologists to denote times of high temperature, sometimes with the idea that the temperature change was rapid. In October 2017, at a conference at the Royal Society in London, geologists, geochemists, palaeontologists, climate modellers and ecologists came together to discuss the growing consensus that there was a common hyperthermal killing model, especially for the end-Permian mass extinction, the Palaeocene–Eocene Thermal Maximum (PETM; a smaller extinction crisis 56 million years ago, see page 219) and the present day.

In the case of the end-Permian mass extinction and the PETM, volcanoes were identified as the driving causes of the crises. Today, volcanoes may not be the main cause of global warming, but the consequences of excess carbon dioxide in the atmosphere are playing out just as we think they did in the deep past. We also know how global warming and ocean acidification can kill life, based on studies of modern plants and animals (see Chapter 5).

How many hyperthermal crises can be identified? During 80 million years of extreme times on Earth, there have been five – the end-Capitanian extinction (259 Ma), the end-Permian mass extinction (252 Ma), the Carnian Pluvial Episode (233–232 Ma), the end-Triassic mass extinction (201 Ma) and the early Toarcian Oceanic Anoxic Event (183 Ma). We have explored several of these in detail, and certainly the end-Permian, Carnian and end-Triassic events were serious catastrophes.

What about the smaller events? By examining the early Toarcian Oceanic Anoxic Event as an example of a smaller-scale hyperthermal crisis, we can explore the components of the model – volcanic eruption, expulsion of greenhouse gases, global warming, acid rain, soil wash-off (discussed in Chapter 6), ocean acidification and seabed anoxia – and determine whether there is something predictable about it. To begin, we head back to the 1840s in Somerset, southwest England, when the first evidence for the event was identified.

MR MOORE FINDS SOME EXCEPTIONAL FOSSILS AT STRAWBERRY BANK

Charles Moore (1815–1881), like so many early contributors to science, was largely self-taught, and certainly was never paid a salary for his scientific endeavours. He was born in Ilminster, a small village in south Somerset, and began his adult life as a printer and bookseller, working in his father's business. At the age of twenty-two, he moved to Bath, where he continued his trade working for the bookseller Mary Meyler & Sons, located close to the Grand Pump Room, a famous attraction for the fashionable people of England, who visited Bath for its natural mineral waters, as described by Jane Austen.

Moore's passion for geology and fossils was stimulated by his early experiences in Ilminster. In the 1840s, he was taking a walk and was surprised by some schoolboys kicking about a rounded boulder. On cracking it open, Moore found a perfect three-dimensionally preserved fish inside. He tracked it back to its source, a quarry above the village, and extracted hundreds more of the nodules, finding fossils in many of them – fishes from a few centimetres to a metre in length, and occasional belemnites, ammonites and other shellfish. That the fossils were preserved in three dimensions was unusual; many of the lovely fossils that Mary Anning had collected along the Dorset coast, especially at Lyme Regis only 29 km (18 miles) south of Ilminster, were flattened – although the marine reptiles were complete, and sometimes showed traces of their skin and body outlines, they had been squashed flat between layers of mud and limestone.

When Moore cracked open the Ilminster nodules, he often saw the bones and shells not only in three dimensions but also in colour. The most common fish, *Pachycormus*, was pretty beefy, up to 1 m (3 ft) long in life, but often only the head and shoulders were preserved. There, in the cream-coloured limestone of a nodule, was a complete

Fossilized head of the fish *Pachycormus* from Strawberry Bank, showing three-dimensional preservation.

fish head, as big as Moore's fist, the bone a light, honey-brown and the skull bones smooth and shiny, as if the fish were lying on a fish-mongers' slab. The mouth was slightly open, showing close rows of teeth along the jaws, and the gills were covered by great bony plates. Behind the head the body was covered with neat rows of rectangular scales, and the fins showed their strengthening rods of bone.

Some of the fossils have soft tissues preserved, such as traces of skin, muscles and guts, and many of the belemnite and vampire squid fossils even have traces of ink. Belemnites are related to modern squid (see page 121) and have sacs of black ink that they can squirt out to confuse predators. Moore knew that geologists had already used this preserved Jurassic-aged ink to write about Jurassic fossils. Most startling of all, Moore found numerous examples of fossil reptiles, ichthyosaurs and long-snouted little crocodiles. Like so many early palaeontologists, he had sharp eyes and even identified hundreds of insect fossils in the limestones around the nodules. When you hold these small, clipped blocks in your hands, it can be difficult to see them, except by turning and twisting them into the light.

Moore kept all the fossils at home, and after he died, the majority ended up forming the basis of the world-famous collections of the Bath Royal Literary and Scientific Institution (BRLSI). The insects, however, are held at the Museum of Somerset in Taunton, where in 2012, for the first time in more than a hundred years, BRLSI Curator Matt Williams and I were able to remove them from the scraps of newspaper and emigration documents from Moore's day in which they were wrapped. Remarkably, Moore never published anything about these amazing discoveries, although he did write many detailed accounts of the geology of his native county, glorying in the fossils he had found. One of the joys of rediscovering and bringing them to wider attention was the name of the fossil site – Strawberry Bank. We scoured early records, newspapers and maps to find its location; the maps showed that a group of houses on the north side of Ilminster High Street were called 'Strawberry Bank', perhaps because wild strawberries once grew on the slope. Up the hill were traces of some old quarries, from which Moore's nodules likely came. Most of Strawberry Bank has been built over, and we did not feel we could ask the owners to dig up their gardens and cellars. However, in an excavation nearby, we relocated the nodule layer in its context in a trench and so are confident of its age.

THE TOARCIAN OCEANIC ANOXIC EVENT

This all matters because what Moore had found were not simply some spectacular fossils, but also the first evidence of the Toarcian Oceanic Anoxic Event (OAE). He had made a lovely drawing of the rock section at Strawberry Bank, indicating just where the insect fossils and nodules occurred, and we can confirm the accuracy of what he showed. His delightfully named 'saurian and fish bed' is associated with ammonites of the so-called *falciferum* Zone.

The allocation to the *falciferum* Zone is a bit of technical detail, but it is important for dating. The unique ammonites of this zone have been identified all over the world, and at one of its other occurrences, in Peru, there are some volcanic ash beds that have provided an exact age date of 183 million years ago. So, even though we don't have such evidence from England, the correlation halfway across the world using ammonites fixes the age everywhere.

Why the ashes in South America? At that time, South America was still fused to the west coast of Africa. Although the North Atlantic was already opening in association with the Central Atlantic Magmatic Province (CAMP) volcanism that drove the end-Triassic mass extinction (see Chapter 9), the South Atlantic opened later. Further, 18 million years after the end-Triassic mass extinction, during the early Toarcian, the Drakensberg Volcanics had begun to erupt in South Africa, spreading their ash widely, peaking at 183 million years ago. Today, these Toarcian-aged basalts form the Drakensberg Mountains, which run along an approximate north–south line in the eastern part of South Africa (see Plate XIII).

The Toarcian OAE is now recognized worldwide, not only in England but also throughout Europe, in South America, South Africa, China and Canada. In many of these locations, the regular deposition of marine sediments is interrupted by black mudstones, indicating oxygen-free conditions on the seabed, and these often contain abundant fossils. During the Toarcian, many communities of organisms experienced severe die-offs and extinctions, but soon life recovered; in fact, it seems there were series of small extinctions, both before and after the early Toarcian peak.

Even though it is called an oceanic anoxic event, it is important to understand that the whole ocean was not stagnant. The anoxia occurred only on the seabed; higher waters contained oxygen and life continued. The seabed is often anoxic on a seasonal scale, flipping back and forwards between summer and winter. In fact, as geologist Rowan Martindale from the University of Texas has shown, the expansion and contraction of the oxygen minimum zone (OMZ) is more of a factor than simply anoxia. The OMZ is a layer of water with low oxygen that does not mix with seawater above or below, and it can rise and fall depending on external conditions. Martindale notes that the Toarcian OAE has been identified in Morocco, although the sediments there show no evidence of anoxia.

The time of greatest anoxia is marked in several places, including Strawberry Bank, by examples of exceptional fossil preservation. It may seem ironic that the wave of killing corresponds to a wave of beautiful fossil preservation, but there were two processes going on. The anoxia enhanced the amount of phosphate in the water and sediment, and this phosphate entered bones and associated soft tissues, ensuring their good preservation in oxygenated conditions. It's what occurred at the Rhaetian bone beds (see Chapter 9), where phosphatization happened in well-oxygenated shallow waters and then the fossils were washed into deeper, oxygen-poor waters.

The same mix of exceptional marine fossils and insects is seen in Germany and Canada. The Holzmaden fossil beds of southern Germany have long been a source of outstanding specimens of ichthyosaurs, plesiosaurs, fishes and other marine beasts. There, German palaeontologists have found hundreds of specimens, often with black-coloured skin outlines; many of the ichthyosaurs are pregnant mothers with numerous embryos inside their rib cages (see Plate xv). Ichthyosaurs were reptiles, but looked like dolphins, and they breathed air. However, they were fully adapted to living at sea, with their torpedo-shaped bodies, a deep sided tail fin to provide propulsion and long paddles for steering (see Plate xvi). The mothers can contain two or three embryos, but sometimes as many as twelve, all lined up like sardines in a tin. It's one of the wonders of Holzmaden that so many pregnant ichthyosaurs have been found, and they are generally at an advanced stage of pregnancy, just before giving birth or even in the process of giving birth. Perhaps the heavily pregnant mothers were at their most vulnerable close to birth, as their embryos had become so large, but it might also say something about the time of year when the Drakensberg volcanoes were erupting. Perhaps the wave of global warming, acidification and anoxia in the shallow seas coincided with the time when ichthyosaur mothers gave birth, probably in spring when food supplies would have become abundant and available for the hungry babies. But was the death of these creatures only as a result of high temperatures, which were as much as 7–13°C (13–23°F) above normal, and ocean anoxia?

EUXINIA: THE SULPHIDE KILLER

An additional factor might have been unusually high levels of sulphur. Sulphur can have profound effects on oceans. When levels are high, the ocean switches to an euxinic state, a condition not seen much today but probably more common in the past. The term 'euxinia' (pronounced 'you-KSIN-ee-ah') derives from the ancient Greek name for the Black Sea, *Euxinios Pontos*, meaning 'Hospitable Sea', because of its rich fishing and generally calm waters, an irony given that euxinic seas, including the Black Sea itself, are extremely inhospitable to life, especially at depth.

Today, deep euxinic waters are known only in the Black Sea and in some Norwegian fjords. Elsewhere, euxinia is seasonal and oceans are generally well oxygenated. The water layers in oceans mix well, with deep waters in the Atlantic and Pacific flowing in cold from the polar seas, then moving towards the Equator, where they rise to the surface, pick up oxygen from the air above and move away from the Equator, to eventually reach the polar waters and descend again. This is the vertical component of the global conveyor belt, the system of great currents that flow through the world's oceans and keep them mixed. The deep-to-shallow-to-deep movement is driven by thermohaline circulation, which depends on the freezing of sea ice near the Poles. As seawater freezes, forming pure-water ice floes and icebergs, the water becomes saltier and denser, and sinks to the seabed, flowing away from the ice cap. This drives the whole cycle right to the Equator, and back again. The horizontal aspects of these greater water flows are driven by the Coriolis effect on the atmosphere and surface waters. As the Earth spins, the atmosphere and surface ocean waters spin clockwise in the northern hemisphere, and counter-clockwise in the southern hemisphere. The Gulf Stream, which flows from the Caribbean, up the east coast of North America and across the Atlantic to northwest Europe, is one example.

But in the Black Sea, there is strong stratification, or layering, of the water, meaning that mixing of waters up and down is rare. The surface waters change temperature with the seasons and contain plenty of oxygen and plenty of life. Deeper than 100 m (330 ft), though, the water is cold and rich in organic matter. This arises because of the Black Sea's unusual situation, land-locked as it is all round, except for the Bosporus Straits between Turkey and Greece that connect it to the Mediterranean. Fresh water flows in from rivers around the Black Sea; the Bosporus enters in the southwest through a narrow channel, and tidal flows mean that brackish water passes south, through the Sea of Marmara to the Mediterranean, at the surface. The reverse, northwards flow is highly saline water, at depth and colder than the surface waters. These saline Black Sea waters remain at depth for up to three hundred years, not mixing at all with the surface waters.

In the presence of abundant organic matter, rich in carbon and trapped in deep waters, as seen in examples with limited water mixing, oxygen levels may already be low. Microbial reactions combine the carbon and oxygen-rich sulphate ions to form carbonic acid and hydrogen sulphide. Hydrogen sulphide (H_2S) is also called swamp gas, sewer gas or stink damp, and most people have smelt the odour of rotten eggs in cold weather coming from pools filled with decaying leaves.

Geologist Hugh Jenkyns at the University of Oxford was one of the first to draw attention to the OAEs of the Mesozoic, and the Toarcian OAE in particular, and their connection to euxinia. He noted the abundant black shales deposited in the oceans during the OAEs and that these black shales were rich in sulphur. Visible evidence for high sulphur levels is iron pyrite (an iron sulphide, FeS_2). Pyrite can form under a variety of low-oxygen, high-sulphur conditions (see page 106), but the golden-coloured cubic crystals are much more common, although much smaller, in muds laid down in euxinic conditions.

Euxinia has been a constant accompaniment of mass extinctions such as the Late Ordovician, Late Devonian and end-Permian crises, as well as several OAEs of Jurassic and Cretaceous age. These were times when the oceans, or some oceans, became sharply euxinic for a short time. Lee Kump, a geochemist at Pennsylvania State University, argued in 2005 that during these times of crisis, the chemocline (the layer separating the deep euxinic waters from surface waters) could rise rapidly to the surface and release huge volumes of hydrogen sulphide into the atmosphere. The trigger for this upward movement can be a combination of an excessive build-up of hydrogen sulphide in the deep waters, perhaps associated with ocean floor anoxia, coupled with lower-than-normal oxygen content in the atmosphere. Hydrogen sulphide upwelling would be fatal to all life, both in the surface waters and living above the oceans and in neighbouring lands. Hence, at Strawberry Bank, perhaps this drove the sudden death of large and small fishes and reptiles, insects and essentially everything in the sea and on nearby land.

DEATH IN THE TOARCIAN OCEAN

Likewise, at Holzmaden, the ichthyosaurs of the early Toarcian were unaware of impending doom. They swam serenely, enjoying the spring-time warming of the surface waters and the arrival of new growth among the plankton and the small shrimps that ate the plankton. Lazily snapping up small fish in their lengthy, tooth-lined jaws, they would flick their heads back and gulp down the shredded flesh. Then something strange happened. The skies had darkened to rich purple and orange caused by the distant huge volcanic eruptions in what is now South Africa. Bubbles of stinking hydrogen sulphide began to shoot up from the depths. The surface waters became more crowded as belemnites, ammonites, fishes and reptiles came up from the deep, driven by the sinister rise of a dark, lifeless ocean layer. Even displaced seabed forms such as lobsters and bivalves were floating uncomfortably at the surface. Then, with a rush from below, the warm waters became shockingly cold and the foul sulphide gases killed everything. Insects plopped into the water like the showers of rice after a wedding.

As the dead fishes, reptiles and other animals floated at the surface and bloated with the gases of decomposition, the high sulphide levels declined. Photosynthesis by plants steadily pumped oxygen into the atmosphere, and the build-up of sulphide at depth had been released. The chemocline descended, as, too, did the carcasses. After a few days under the Early Jurassic sun, the ichthyosaur and fish carcasses surfaced and exploded, and the unwieldy skeletons with ripped flesh sank slowly into the deeps, where in the black, oxygen-free, sulphur-rich muds they were not perturbed by scavengers. Some were buried quickly, preserving hard and soft tissues. At Strawberry Bank, but not at Holzmaden, the fossils were covered by muds, and thanks to further chemical reactions, began to convert it into hard stony shells of many layers, which would later become nodules, to be kicked by unappreciative schoolboys or cracked open by palaeontologists.

The Toarcian OAE is not usually identified as a mass extinction. There were turnovers among ammonites, bivalves, belemnites and brachiopods, with many extinctions of species, and many new species appearing. Most dramatically, there was a major reef crisis that caused substantial extinctions and a complete remodelling of reef systems. None of the ichthyosaurs or fishes that died so dramatically went entirely extinct. Perhaps the chemocline rise, and the flooding of the atmosphere with hydrogen sulphide, happened only at certain spots, depending on ocean depth, climate and latitude.

On land, the dinosaurs were evolving and changing: early forms gave way to new groups such as larger predatory theropods, the colossal, long-necked plant-eating sauropods and new kinds of armoured dinosaurs, such as ankylosaurs with all-over body armour and stegosaurs with spines and plates along their backs. Lizards, crocodiles and mammals were all present and evolving, but there is no sign of any major or sudden extinction.

If euxinia was a key killer in the Toarcian OAE, how does that connect with the other elements of the hyperthermal model?

THE HYPERTHERMAL MODEL

Now is the time to try to draw the various lines of evidence together. It all begins with a volcanic eruption. As noted earlier, when we think of volcanoes, we think of pointy-topped Vesuvius and Etna in Italy or Mount St. Helens in North America. These are called Plinian volcanoes after Pliny the Younger (AD 61–c. 113), who in AD 79 witnessed the famous eruption of Vesuvius and provided the first written account of what happened. So began volcanology as a science. In the case of the great mass extinctions, however, the volcanoes involved were nearly always shield volcanoes, which poured their lava out through fissures, or cracks, in the Earth's crust, but did not develop mountainous piles of cinders and lava. Such shield volcanoes may be seen in Hawaii and Iceland where they tap into huge magma chambers beneath the oceans. In Iceland, fissures traverse the mid-ocean ridge, the site of a steady upwelling of lava that forms the floor of the Atlantic Ocean, the modern manifestation of the CAMP eruptions at the end of the Triassic when the North Atlantic Ocean was but a dream.

Whatever the volcano type, eruptions generate not only lava but also gases that are squirted high into the atmosphere. These gases are of many kinds, the most abundant of which is water vapour, then carbon dioxide, sulphur dioxide, hydrogen sulphide and hydrogen halides. Methane can also be produced when the lava flows burn through organic matter like trees.

Water vapour, carbon dioxide and methane are all so-called 'greenhouse gases' that rise and then shield the reflected Sun's rays from escaping from the atmosphere, so warming the Earth.

Eruption of Eyjafjallajökull in 2010, showing how a slowly erupting basaltic volcano can generate huge amounts of gases that affect a very wide area.

Sulphur dioxide also rises through the atmosphere, but at height has an opposite effect to the greenhouse gases, forming a layer that reflects the Sun's rays and so can cause cooling. This effect is transient, however, and quickly overwhelmed by the greenhouse effect. The hydrogen halides, such as hydrogen fluoride, can break down the ozone layer (which absorbs most of the Sun's harmful ultraviolet rays) at the top of the atmosphere, allowing excess ultraviolet rays to reach the surface of the Earth and cause harmful genetic mutations in plants and animals. When mixed with water in the atmosphere, all these gases convert to acids – for example, sulphur dioxide becomes sulphuric acid (the acid in car batteries). Hydrogen fluoride eats away rocks and glass – it is one of the worst chemicals to handle in terms of laboratory safety; if poured into a test tube, the test tube will simply melt. As we have seen, the gases from volcanic eruptions generate both sharp global warming and acid rain, coupled with ocean acidification.

How do we know this? First, of course, all the chemical reactions just mentioned have been understood for centuries and are taught as basic Chemistry. Second, volcanologists study modern volcanic eruptions with enthusiasm. It is known that these eruptions are killers, not only because of the lava and falling ash (as in Pompeii in AD 79), but also through gas release – there are low-lying areas along the African Rift Valley where people and farm animals die mysteriously with no indication of the killer. These deaths are caused by carbon dioxide leaking from volcanic cracks and building up silently in low-lying areas. The gas kills and can then blow away, leaving no trace.

Studies of modern volcanoes also provide the scaling factors needed to understand huge volcanic eruptions of the past. For example, the eruption of Mount Pinatubo in the Philippines in 1991 injected 250 million tonnes of carbon dioxide and 17 million tonnes of sulphur dioxide into the atmosphere. The atmospheric effects of cooling, then warming, produced by the eruption were detected worldwide. Krakatoa, which erupted in Indonesia in

1883, had similar effects. Locally, the explosion and debris killed 36,000 people. Debris was thrown 40 km (25 miles) into the air and the explosions were heard as far away as Australia, 4,500 km (2,800 miles) distant. The skies were darkened by the debris and dust up to 440 km (275 miles) distant, and ash fell on the decks of ships 6,000 km (3,770 miles) away. The atmospheric effects of the sulphur dioxide were detectable worldwide, with colour changes to the skies over Europe and North America, and temperatures were 1.2°C (2.2°F) cooler for the next five years. These kinds of studies show that volumes of lava and volumes of gases are predictable based on the size of the eruption.

The erupted volcanic gases have an uncanny resemblance to the polluting gases that humans are today releasing into the atmosphere by burning coal and other fossil fuels. The darkness and cooling from sulphur dioxide may be transient but is nonetheless serious. We are already seeing global warming, acid rain and ocean acidification (see Chapter 15). These are the predictable consequences of the hyperthermal model – it doesn't matter whether the carbon dioxide comes from gas-guzzling cars or a volcano. The killing of life on land by heat and acid rain, and life in the sea by heat, acidification, anoxia and euxinia, have always been consequences of massive eruptions (see Chapter 6). The emergence of the common hyperthermal model in the past twenty years has been a profound outcome of intensive efforts to understand the modern global crisis, but especially the scaling of extinction events in the past.

Following the Toarcian OAE in the Jurassic, there were further such events in the following Cretaceous before the next mass extinction, the famous end-Cretaceous crisis when the dinosaurs disappeared (see Chapter 12). But, in the mid-Cretaceous, something was stirring in the undergrowth – new kinds of plants, insects and insect-eaters. In fact, a quiet revolution transformed the surface of the land in the Cretaceous and after the dinosaurs had gone, and this was probably much more significant for modern biodiversity than their disappearance.

PART 5

The Cretaceous to the End of the Eocene

145–34 million years ago

CHAPTER 11

The Angiosperm Terrestrial Revolution

THE STEAMING TROPICS OF MYANMAR

It's midsummer in Myanmar (formerly Burma) 100 million years ago. The forest by the sea is made from tall monkey puzzle and cypress trees, with smaller flowering plants between them – laurels, dogwoods and woody bushes with rose-like flowers. Ferns occupy damp spaces on the ground, and the branches of the trees are covered with mosses and liverworts. Sticky resin oozes from cracks in the bark of the conifers or from broken branches.

There's a hot, sharp, high-pitched background hum that never stops, like little buzz-saws, first one, then another. Occasionally, a hefty scarab beetle flies past, its beautiful black, patterned carapace flashing in the dappled sun, rocking uncertainly on its small wings. Huge numbers of tiny flies mill and dance in the rays of the Sun. A small moth with striped wings flutters by. Hard little wasps, striped yellow and black, zip purposefully from flower to flower, seeking nectar. On the branches, millipedes and spiders, some rather large, move tentatively and scuttle into cracks in the tree bark.

A gecko lizard moves along a narrow branch, testing each foothold, the rest of its body swaying in anticipation, and then it places its foot purposefully, still vibrating slightly, and raises the other foot. The spiders and insects seem oblivious to the advancing gecko, which pauses, cocks its head in thought, then flashes out its sticky tongue to snatch a cricket as its song reaches a crescendo.

There's a nest in one tree, a rough construction of twigs and moss, and there are three baby enantiornithines inside. These are fluffy little nestlings like any modern tree-dwelling bird, although not closely related to any particular one. They only hatched three days ago, and their eyes are still closed. They are mainly naked, pink and bony, almost like anatomical preparations, their eyeballs showing black beneath the skin, their internal organs pulsing inside their narrow chests and a soft fuzz of feathers over their head and shoulders. They sit in the nest, squirming, extending their bony winglets and prodding their siblings without mercy.

An adult enantiornithine approaches the nest, braking slightly in flight and approaching the branch nervously; it lands with a flurry of feathers and hoarse squawks, then offers the grub in his mouth to the insistent babies in the nest. One that is larger than the others lurches forward expectantly, snatching for the tempting grub, elbowing his brother and sister in doing so. He teeters forward in the nest and loses his balance, tumbling and rolling down the trunk of the great cypress tree. The small branches slow his tumble, and his little feathers get caught on snags in the tree bark, gaining blobs of gummy resin.

The bird comes to a halt, 15 m (50 ft) below the nest and a similar distance above the forest floor. His father sees him and swoops down, chittering angrily, but he can do nothing. The nestling is stuck, its feet covered in resin, and he will have to save himself. He flaps his little wings, each the size of a postage stamp and bearing only a dozen or so tiny feathers that are nowhere near ready for flight. The baby wrestles but cannot free himself, in any case now too weak to crawl back up the tree to the nest. As night falls, the air becomes colder and eventually wet. He shivers, grumbling and hungry. Before dawn, he has died.

Over the course of a few months more resin flows from the tree and the baby bird is partly encased, one wing being entirely covered. Then, a few years later, a great storm hits the forest and trees tumble, including the cypress with the entrapped baby bird. In the next days and weeks, further storms tumble the tree trunks, breaking off branches and dislodging blobs of resin with their trapped insects, spiders and the baby bird. River flows carry these fragments a short distance out to sea where they are deposited in beds of fine-grained sandstone. After burial, the resin hardens and eventually turns into amber, a glassy semi-precious stone that retains the flow patterns of the original resin, including any trapped twigs, seeds, insects or other organic fragments inside. Without exposure to the air, these do not decay.

One hundred million years later, the amber deposits of northern Myanmar are mined by local citizens who earn their living by selling amber for necklaces and pendants. Nicely shaped pieces of amber that contain an insect are especially prized and fetch a higher price. Then, over the past hundred years, palaeontologists from around the world have homed in on the amber markets on the borders of Myanmar and China, where certain pieces with unusual fossils, like frogs, lizards, isolated feathers or even, as in this case, remains of a tiny bird wing, realize the highest prices of all. This particular example was purchased by Lida Xing, a palaeontologist from Beijing, who studied it in detail using X-ray scanning to see inside the amber, and I was invited to be a co-author of the paper in which the specimen is described, eventually published in 2016.

Burmese amber is very special – and very controversial. The ongoing civil war in northern Myanmar has cast a long shadow, and it is feared that some or most of the mining and sales may involve semi-slave labour and raise money that fuels the armed conflict. Many palaeontologists have implemented bans on scientific publications on amber that reached the market after either 2017 or 2021. But from the research done so far, it is evident that

Exceptionally preserved flowers in Burmese amber, with five spreading petals, like the modern Cunoniaceae from Australia.

Burmese amber is immensely rich in fossil diversity, with over a thousand insects and other creepy-crawlies named so far, representing 44 families of spiders, over 500 families of insects, 25 families of plants and 20 or more species of vertebrates. The world it portrays is very different from our impression of life in the Mesozoic. It is not only that the emphasis is on tiny animals caught in amber rather than the great skeletons of dinosaurs in ancient river sands; we also seem to be witnessing a whole new world – a revolution, something bigger than the usual stop-start steps of evolution.

INTIMATIONS OF A REVOLUTION

The burgeoning scene of 100 million years ago provides a picture of the beginning of the Angiosperm Terrestrial Revolution (ATR for short), a term I coined in a paper in 2022 with palaeobotanist Peter Wilf of the Pennsylvania State University and evolutionary botanist Hervé Sauquet of the Royal Botanic Gardens in Sydney,

Australia. In this chapter, we will explore the role of the ATR and how it relates to the end-Cretaceous mass extinction.

Until recently, many biologists tracked the origins of modern groups of plants and animals back to the time when life recovered from the end-Cretaceous mass extinction, 66 million years ago (see Chapter 13). It's commonly understood, for example, that mammals got their chance to diversify only after the dinosaurs had lumbered off into the sunset. This surely marked the beginning of the modern world, with its singing birds, bright flowers, busy bees and butterflies, and everything we appreciate about nature? Well, not so. All those 'modern' groups of plants and animals can be tracked back to the middle of the Cretaceous, 100 million years ago.

In some senses, the end-Cretaceous mass extinction was a punctuation in a process of diversification on land that we witness in the Burmese ambers. The flowering plants (angiosperms) are already present, as are modern types of pollinating and social insects such as ants and wasps, as well as the lizards, birds and mammals that prey on them. In our 2022 paper, however, we noted that the ATR had two phases: one in the mid-Cretaceous, and one in the Palaeogene, the later recovery time after the demise of the dinosaurs. The first phase, discussed in this chapter, marks the origins of all the major groups; the second, discussed in Chapter 13, marks their explosive expansion.

The story starts in 1980, when a young Danish palaeobotanist Else Marie Friis took up a British Council Research Scholarship in London. She had earned her PhD that year working on fossil plants from the lignite (soft brown coal) mines of Central Jutland in Denmark, only some 15 million years old. The mines were, of course, economically important, but there she learned how to extract and study beautifully preserved fossil leaves, flowers and seeds from flowering plants. Dr Friis switched topics for her London fellowship, moving back in time to the Cretaceous of southern Sweden, where she and Swedish scientist Annie Skarby had discovered

Exceptionally preserved fossil flowers of the daisy-like *Bertilanthus* from the Cretaceous of Sweden; each flower is about 2 mm long.

remarkable fossilized flowers when sieving through fossil-rich sediments from a quarry in Scania. Among the leaves, twigs and seeds, they found tiny flowers, only 2 mm long. The new Cretaceous flowers, later called *Bertilanthus*, showed all the petals, sepals, anthers, pollen, stigma and other key parts of a modern flower. Friis and Skarby assigned their flowers to Saxifragales, a modern order of 2,500 species of witch hazel, currants, lianas and peonies.

Until then, all that was known of Cretaceous angiosperms were fossil pollen and leaves and some incomplete flowers, so this was a hugely important discovery because it confirmed that Cretaceous angiosperms had already evolved their key feature:

the flower, each with its symmetrical whorls of petals and sepals, as well as the male and female sexual organs at the centre. As every biology student learns, the new generations of plants form as the ovules, which are contained at the centre of the flower in a fleshy bottle-like structure called the pistil. The slender male structures, or stamens, are positioned around the pistil, each tipped with an anther that produces pollen. The pollen, containing the sperm, is transferred by wind or animal intervention to the upper end of the pistil, called the stigma, where each sperm constructs a tunnel down to the ovules, where fertilization occurs. The ovules then develop into seeds.

In angiosperms, the remarkable specialization of double fertilization is often said to be vital to their success. Rather than one male pollen grain uniting with one female ovule, two sperm are involved in fertilization. One unites with the egg nucleus, while the other fuses with another nucleus that divides to form the food supply, called the endosperm, for the developing embryo. Therefore, each fertilized ovule becomes part of a seed accompanied by the endosperm. The production of a seed with its own food supply guarantees that it will have a good chance of survival. When a seed is planted on damp paper, the root pops out one way, the shoot the other, and for some days or weeks the fledgling plant grows happily, feeding from the endosperm. In nature, the seed would also extract water and additional mineral food from the soil.

But seeds – in the form of fruits, nuts and grains – with their own supply of nutritious food, are also very attractive to animals. Squirrels eat nuts, humans eat fruit, peas and beans, as well as grass seeds such as oats and wheat. The same is true of all the nectar (sugar water) that is produced in flowers, which is slurped happily by hordes of small creatures. Surely fruits, nuts and grains are bad news for angiosperms if they are eaten by animals, having expended so much energy in producing them? In fact, the exact opposite could be argued: that animals are the slaves of the angiosperms.

FLOWER POWER

Flowers exert their power over animals in both pollination and germination. The beautiful flowers we appreciate so much exist mainly to lure in insects, such as bees, moths and butterflies, which are attracted to a flower by its shape, colours and scent. They dive in, seeking the nutritious pollen and nectar produced by the plant, which they slurp up through specialized mouth parts. When entering and leaving, they brush against the flower's anthers and pick up pollen, then fly off to the next flower, where they deposit the pollen, and fertilization can begin. The exquisite co-adaptation of certain flowers to their pollinators is astonishing – for example, certain hummingbirds are adapted to particular deep, tubular flowers that precisely fit their beaks. Coevolution means simply 'evolution together', and such exquisite cases suggest millions of years of close coexistence between flower and hummingbird.

Today, about 24,000 of the 300,000 species of angiosperms (8%) are wind pollinated, and the remaining 92% are pollinated by animals – usually quite specific groups of insects, birds or bats. For example, bats and moths come out at night so the flowers pollinated by these open and release scent at night. Coevolving flowers and pollinators have matching histories and geographic distributions, and some astonishing fossils show insects covered in pollen – a real example of being caught in the act!

So, it's no surprise then that fruits, nuts and grains have evolved specifically to be eaten. After consuming the nut or fruit, the animal moves around and then when it defecates sometime later in a different spot, the seeds come out with a strongly nourishing dollop of manure in which they happily sprout and grow. If this did not happen, and the seeds all fell straight to the ground, the whole area would be choked with baby plants of the same parentage. To produce a fruit, have an animal eat it and plant it, then compost it some distance away is a smart evolutionary strategy. Animals are mere automata, and in this scenario are the slaves of the plants with which they have coevolved. We break the chain when we fastidiously remove orange pips and apple cores or cook our pulses and grains, so they do not go on to germinate, but in general, the angiosperms are getting exactly what they need in evolutionary terms.

Flower coevolution was one of the fundamental features of evolution that fascinated Charles Darwin. He famously described the origin of flowering plants as an 'abominable mystery', meaning that their adaptations and co-adaptations with pollinators and seed-eaters were so exquisite that in many cases he could not see how this complexity had evolved. Now, thanks to astonishing fossil discoveries, we can see the transition from non-angiosperm to angiosperm as it happened in the Cretaceous. In fact, all the evidence from the shape of the evolutionary tree is that angiosperms evolved much earlier. There are many reports of possible Triassic and Jurassic ancestors, although these have mainly been rejected and we must keep hunting for the elusive first flowers. It is also likely that many of the specialized features of flowers occurred earlier in related groups. Whatever the case, the early angiosperms evidently did not gain much of a toehold on life and only began to take over ecosystems in the mid-Cretaceous.

Some of the coevolution we see so finely tuned today might have been a little less so in the Cretaceous – for example, particular angiosperm species might have been open to all pollinators and perhaps their intimate coevolved associations emerged later. There are also cases where Cretaceous ancestors of modern angiosperm species favoured different pollinators than they do today. Organisms adapt to circumstances, and one beetle group or another that served well as a pollinator 90 million years ago might have gone extinct, moved or evolved in such a way that another group took over and the plant evolved to accommodate its new partners.

These recent discoveries about early angiosperm evolution have massively illuminated Darwin's abominable mystery. Friis, who reported the first well-preserved Cretaceous fossil flowers in 1981, has continued her work, describing new specimens from localities in North America, and more recently from China. She has written definitive books on the topic and been widely honoured not only in her native Sweden, but also in Denmark, Norway, China, Britain and the United States.

HOW DOES A BIRD BECOME A BIRD?

Biologists love successful groups of organisms, which is why they lavish such huge attention on angiosperms. Birds hold the same attraction: there are ten thousand species and they are everywhere. Success is a great theme in evolutionary biology: why are some groups, such as angiosperms and birds, and even more so the insects, so species rich? The usual answer is that they each evolved some very special set of adaptations, or innovations, that enabled them to conquer a whole new mode of life. In the case of birds, it is all the features that enabled flight (small body size, wings, feathers, enhanced senses) and, having evolved that package of features, birds took off, so to speak, and reached their current high level of biodiversity.

It is this level of success that raised a big question for Darwin, and indeed for critics of evolution. How exactly could a bird – or an angiosperm – first evolve? Each is hugely good at what it does, as evidenced by its diversity and abundance, and each represents such a package of exquisite adaptations to its unique mode of life, that it seems impossible to imagine how they could have evolved through normal processes. Some evolutionists in the twentieth century despaired of finding the answer, even suggesting special kinds of accelerated evolution. After all, what use is half a bird? We now know that they were wrong to despair and that half a bird is, in fact, fine. The problem was simply a failure of imagination.

Even more than the angiosperms, there is now a rich fossil record of the numerous steps between dinosaurs and the first bird, and the first bird and modern birds. The enantiornithine that was trapped in the amber of the Myanmar rainforest 100 million years ago shows a cluster of features midway between the oldest bird and modern birds. Other fossils, especially from China, show us how birds became birds in a piecemeal fashion.

When I was a student, we spent a lot of time studying *Archaeopteryx*, or Urvogel ('primeval bird'), the famous oldest bird from Germany. There are a dozen specimens of this remarkable beast from the latest Jurassic, about 150 million years ago, showing its skeleton and feathers in some detail. It's a bird and it could fly, but it retained many reptilian features such as a long bony tail, an unreinforced pelvis and teeth in its jaws. We counted about thirty uniquely bird-like features that appeared first in *Archaeopteryx*, including expanded eyes and brain for excellent vision, wings, feathers, a specialized wrist that let it fold its wings back, a fused clavicle or wishbone, hollow bones and many more. This was the question: did *Archaeopteryx* spring onto the scene with all these flight features, or could we imagine a long series of ancestors with just a few of the full set?

The answer is, yes we could. Since 1996, Chinese palaeontologists have delivered a flock of amazing little feathered creatures to a world agog (see Plate 11). After the very unexpected discovery that many dinosaurs had feathers was accepted, Darwin's conundrum began to unravel and rebuild in a rational manner. Thanks to the discovery of thousands of specimens of a hundred or so new species from China, it is now known that the first of those thirty avian features of *Archaeopteryx* had emerged over 50 million years earlier in the Late Triassic, and through the Jurassic other features were acquired piecemeal, so by the time *Archaeopteryx* appears in the evolutionary tree, all the unique bird features are present.

At first, feathers were for insulation and display, not for flight. Hollow bones save weight, but they are part of the respiratory system of many dinosaurs, too, accommodating air sacs connected to the lungs to improve the efficiency of breathing. Even flight feathers themselves were not at first used for powered flight (flapping). Today, many vertebrates other than birds and bats can fly

– there are flying fish, flying frogs, flying lizards and snakes, and numerous flying squirrels and other mammals. Of course, they fly by gliding, but that serves them well. Likewise, in the Jurassic, a whole panoply of theropods miniaturized and probably began to occupy the trees, seeking out insects, lizards and other small prey. Expanded arms with proto-wings were a great aid to these small predators when they leapt after their insect prey – even a small wing could extend a jump by a few metres.

New research shows that four or five lineages of theropod dinosaurs crossed the threshold between gliding and flapping at different times in the Late Jurassic and Early Cretaceous. Some of these early experiments with powered flight involved two wings like birds; others had four wings, on arms and legs; yet others had membrane wings like bats. Ultimately, *Archaeopteryx* represents the successful line leading to all later birds, including our unfortunate fledgling in Myanmar. The other gliders and flyers eventually died out without issue. The fossils have shown, though, that despite claims to the contrary, half a bird is fine; even one-tenth of a bird can be a perfectly well-adapted creature.

The same argument is true of angiosperms. For example, insect pollination almost certainly began before the origin of angiosperms, so they took over an existing coevolutionary model, and pre-angiosperms might not have been bisexual (that is, including both male and female reproductive structures). Probably half an angiosperm functioned perfectly well, although there is not a great deal of evidence about how the angiosperm flower was assembled through stepwise evolution.

HOTHOUSE WORLD

The Late Cretaceous was a time of considerable warmth and high sea levels that enabled angiosperms to flourish. At about the time of the Myanmar rainforests 100 million years ago, a worldwide phase of plate tectonic activity took place. Magma welled up from below the mid-ocean ridges and new crust began to be formed. As the magma pushed up, the floor of the oceans rose and water overflowed, so that sea levels rose by as much as 200 m (660 ft), and coastlines shrank as all the land up to this level flooded. Great seaways bisected Africa and North America. At the same time, all the active magma eruption pumped carbon dioxide into the atmosphere, producing a long phase of warming.

Although there were some small-scale extinctions, these did not seem to greatly affect the angiosperms. These were oceanic anoxic events (OAEs), like the Toarcian event at Strawberry Bank in Somerset (see Chapter 10). The first (traditionally called OAE1) occurred 117–116 million years ago, and the second (OAE2) at 94 million years ago. Both appear to have been hyperthermal events, perhaps driven by eruptions in the southern Indian Ocean and the Caribbean respectively, with all the consequent gas ejection, warming and acidification. Both caused extinctions that were not at a huge scale and in part regional. OAE2 did see the end of the ichthyosaurs, the dolphin-shaped marine reptiles that had been so important in Jurassic and Cretaceous seas (see Chapter 10). The impact of these events on plant evolution is less certain, but a cold event that formed part of OAE2 was marked by the temporary retreat of forests and expansion of open, savannah-type vegetation that favoured the angiosperms. The phase of high temperatures and high sea levels did not last, however, and temperatures began to fall through the last 20 million years of the Cretaceous. This marked the subsidence of the ocean floors as plate tectonic activity reduced and sea levels began to fall. Greenhouse gases from the mid-ocean ridge volcanics fell back, so temperatures began to cool.

During the Late Cretaceous, in general, angiosperms were on the march. They did not take over the world rapidly, but comparisons of sequences of floras show how they steadily became an increasingly larger proportion of the species counts. Starting at 0% of typical floras in the Early Cretaceous, they rose to 5% in ten million years, then 20% after 30 million years, 50% at the time of OAE2 and 75% by the end of the Cretaceous. This rise was partly at the expense of pre-existing groups such as conifers and ferns, but much of it was a straight addition. In other words, although angiosperms comprised 75% of the latest Cretaceous floras, the other plant groups had only halved in diversity. Angiosperms clearly had a capacity to outcompete other plants, but also to add new ecological forms. Latest Cretaceous floras contained twice as many species on average as the Early Cretaceous floras.

Strong reasons for angiosperm success can be found in their stomata and leaf veins. Leaves have one main function, which is to capture sunlight. As we know, typically plants photosynthesize, a process that combines carbon dioxide from the atmosphere with water, from the ground below and from the atmosphere, to produce oxygen and carbon-rich sugar. Photosynthesis is driven by the energy of the Sun, which is why most plants grow more on a sunny day. The carbon dioxide passes into the leaf and oxygen passes out through remarkable small openings under the leaf called stomata (singular 'stoma', meaning 'mouth'). In angiosperms, there are twice to four times as many stomata as in other plants. Angiosperms also have up to five times as many veins in their leaves (see Plate XXIV). Veins carry water sucked up from the ground through the leaves and gas exchange happens through the stomata, meaning an angiosperm leaf is up to five times more efficient than leaves of other plants. The result is that a patch of angiosperm forest can capture more of the Sun's energy than other plants can.

We have also seen that angiosperms could simply double the number of species in a patch by double fertilization, reflecting their much greater ability to adapt and take on varied ecological roles. By the end of the Cretaceous, angiosperms were diverse, with all kinds of flowers, but the plants were generally bushy in height and mostly did not yet dominate as trees, roles that were still filled in the landscape by conifers and other gymnosperms. So, the Angiosperm Terrestrial Revolution had begun, but it was by no means finished. It would take a dramatic mass extinction and some climate change to see the second even more impressive phase that completely transformed modern ecosystems and made modern biodiversity so high.

The latest Cretaceous on land was a world both familiar and unfamiliar (see Plate XVII). In some ways it looked modern, with forests full of frogs, lizards, snakes, small birds, mammals and flowering plants, such as roses, lilies, laurels and dogwoods, with beetles, flies, butterflies, and bees buzzing around. Dominating all these, though, were the dinosaurs, classic beasts such as *Tyrannosaurus rex*, *Triceratops* and *Ankylosaurus*, and not at all of the modern world. These latest Cretaceous worlds are known in a great deal of detail, even those virtually on the last day of the Cretaceous when the most famous mass extinction of all was driven by the most unexpected of catastrophes.

CHAPTER 12

The Day the Dinosaurs Died

TANIS, NORTH DAKOTA

The Tanis fossil site is just within the southern border of North Dakota in the USA, near the town of Bowman, a community of 1,700 people on Highway 12, which meanders for hundreds of miles across hot, open prairie. The site was discovered in 2008, but came to fame after 2019, when palaeontologist Robert DePalma, then a graduate student at the University of Kansas, began to make some remarkable claims. He had found evidence, he said, that the Tanis site exposed a layer in the rock that corresponded to the exact day that a huge asteroid struck the Earth. The fossils at the site, including fishes, freshwater turtles, dinosaurs, pterosaurs and mammals, had been killed by the shock waves and fallout of that asteroid strike. Further, he claimed in 2021 that the impact happened in late spring, although the year of impact was not known.

DePalma's discoveries at Tanis were reported widely in the press, including major news stories in *The New Yorker* and in *Science* in 2019. This was followed by a further burst of attention in early 2022 after the BBC aired *Dinosaurs: The Final Day* with David Attenborough in the UK, retitled *Dinosaur Apocalypse* in North America. In the documentary, many scenes from the Tanis site are seen. The palaeontologists drive out over endless flat badlands on rough roads to reach their destination. In summer, the grass dies back and the landscape looks like a semi-desert. The excavators have to set up awnings of poles and canvas to protect themselves and the fossils from the burning rays of the Sun. At times, heavy rains hit the open ground and deep coulees (gullies) are eroded by the roiling torrents. In summer, these watercourses are dry, but they cut up the land and are the main reason this kind of landscape has been called 'badlands' – bad for farming, bad for horse-riding, bad for everything except rattlesnakes and dinosaurs.

I experienced some of the to-and-fro of debate during 2021 and early 2022 when I was scientific advisor for the documentary and worked with the film-makers, discussing in detail what they planned to show. When the programme aired around the world, it brought intense attention on this very special site in North Dakota. So, exactly what are the claims and the evidence?

THE LAST DAYS IN THE HELL CREEK FORMATION

The first claim about Tanis is perhaps the least controversial, namely that the site documents the very end of the Cretaceous, the years or months before and right up to the impact. Tanis sits within the Hell Creek Formation, a flat-lying packet of 50–100-m (165–330-ft)-thick rock that covers thousands of square miles of Montana, North Dakota, South Dakota and Wyoming. The rocks are mudstones and sandstones laid down by ancient river systems that spread over the whole area in the Late Cretaceous at a time when global sea levels were up to 200 m (660 ft) higher than they are today (see page 186). In fact, the continent of North America was divided in two by the Western Interior Seaway, which formed a tongue of ocean 970 km (600 miles) wide running north from the Caribbean through Texas, Colorado, Wyoming, the Dakotas,

Alberta and the Northwest Territories of Canada, before reaching the Arctic Ocean. The Hell Creek rivers flowed mainly eastwards towards this ocean.

Fossils have been collected in the Hell Creek Formation for over 150 years; they include rich floras of diverse tropical-climate plants, populated by abundant insects and snails, and dinosaurs such as *Tyrannosaurus rex*, *Triceratops*, *Edmontosaurus* and *Euoplocephalus*. The lakes and rivers were occupied by a wide variety of fishes, frogs, salamanders, crocodiles and turtles. Early mammals scuttled in the undergrowth and pterosaurs and birds flew in the air. Many of the classic dinosaurian stand-offs in art and films between *T. rex* and his many foes and victims have been based on fossils from this area.

During the 1980s and 1990s, palaeontologists carried out detailed censusing studies of fossils in the Hell Creek Formation, documenting the exact levels of each fossil. The idea was to determine whether the dinosaurs had died out with a whimper or a

The Hell Creek Formation, Makoshika State Park, Montana. This massive geological formation includes the Tanis site in North Dakota that records the last day of the Cretaceous, the day the dinosaurs died.

bang; in other words, had they survived in full force to the very end of the Cretaceous or did they decline through many metres of rock strata below the extinction layer? The consensus was that they survived pretty well until nearly the end, but then apparently no dinosaur bones occurred in the top 3 m (10 ft) of latest Cretaceous rocks. This 'dinosaur gap' was interpreted variously to mean either that the dinosaurs had died out some months or years before the asteroid impact or that for some reason their bones weren't being found.

How do we identify the end of the Cretaceous in these Hell Creek sediments? For a long time, geologists had no exact tools, so they used a widespread coal layer called the 'Z (or 'zee') Coal' as the marker of the end of the Cretaceous period and the beginning of the subsequent Palaeogene period. This worked well enough because the coal was easily identified in the field, and it did mark the end of the dinosaurs – dinosaur bones were found below the coal, but never above. The level was confirmed as close to the Cretaceous–Palaeogene boundary by a change in pollen types in the sediments, and the extinction of a broad range of typical Cretaceous plants just below the Z Coal.

Even so, this is not enormously accurate, and it doesn't work everywhere. For example, at Tanis there is no sign of the Z Coal, and the only field marker for the end of the Cretaceous is the disappearance of the last dinosaur. But using the absence of dinosaur bones to mark the end of the Cretaceous isn't a great way to date the rocks if it's the end of the dinosaurs we are interested in. This is classical circular reasoning – 'Here's the end of the Cretaceous marked by the last dinosaur bones in the rocks; oh, and this shows us how the dinosaurs declined in the last days of the Cretaceous.'

There's now an independent age marker for the end of the Cretaceous, and this is a thin layer of clay containing elevated proportions of iridium. Iridium is a metallic element, related chemically to platinum and gold, and even rarer than they are in the Earth's crust. In fact, iridium is generally assumed not to occur

(XVII) Bucolic scene in the Late Cretaceous. Yellow flowers and tall magnolia trees make the scene look modern, but the large pterosaur reminds us this was in fact 70 million years ago. The angiosperms (flowering plants) that form the framework of modern terrestrial ecosystems were already there.

(XVIII) End-Cretaceous impact. The most famous and most shocking crisis on Earth happened 66 million years ago, when a huge rocky meteorite hit our planet, forming a great crater in southeast Mexico. The backblast from the impact caused worldwide dark and cold.

(XIX) Meteor Crater, Arizona, USA. The name says it all. This crater, 1.2 km (3,900 ft) across, shows the key features. The circular shape is caused by the backblast as the Earth's crust reacts to the impact, throwing debris high in the air, creating a lip round the edge.

XX Shocked quartz. Quartz is the most common mineral in rocks and is usually unlined. Under high pressure, numerous sets of lamellae (layers) form, here at least two that cross at slightly different angles. This example comes from the time of the end-Cretaceous impact.

XXI Unhappy pterosaurs. Here some giant *Pteranodon* meet the molten debris of the backblast from the end-Cretaceous impact. The bombs are debris thrown up by the backblast of the impact coupled with a heated shock wave. These pterosaurs are not long for the world of the living.

XXII The famous chalk cliffs of Dover. In the Late Cretaceous, sea levels were up to 200 m (650 ft) higher than today, and warm oceans were filled with microscopic plankton that rained down on the seabed forming great thicknesses of chalk.

XXIII Coccolithophores under the microscope. Coccolithophores, plants that photosynthesize at the surface of the ocean, were the key plankton of the Late Cretaceous that formed chalk on the seabed. In life, they are protected by exquisitely beautiful circular plates of calcium carbonate.

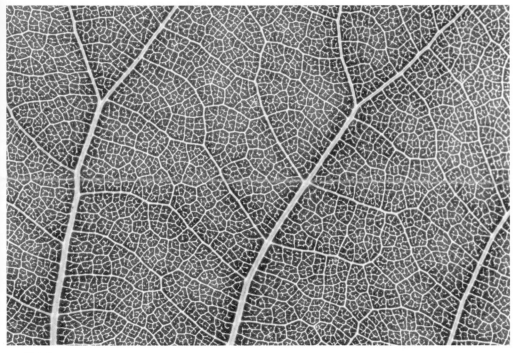

XXIV Secret of the success of the flowering plants. One of the key features of angiosperms is that they can photosynthesize faster than other plants. This is partly enabled by the efficient distribution of leaf veins that transport fluids around the whole plant.

XXV Weird world of the Palaeogene. Flowering plants expanded over tropical zones as rainforests, a new kind of habitat. In places, birds became flightless, like these *Gastornis* or *Diatryma* who lived in Europe and North America, and were in places the dominant large animals of their day.

XXVI Ice age giants. Here, a herd of mammoths walks over the thawing ground in spring as tundra vegetation begins to flourish. In the foreground two cave lions watch over their reindeer prey, and a woolly rhinoceros on the right looks over to two wild horses, at far left.

XXVII The polluted world, part 1. This satellite photograph of North America shows the extent of light pollution in the middle of the night. Cities, public highways and factories are lit using power from fossil fuels. Each night, thousands of tonnes of coal, or equivalent, power the lights.

XXVIII The polluted world, part 2. Plastic debris floating in the ocean should not be there, whereas the jellyfish, lower right, should. Jellyfish absorb plastics as small particles, whereas other sea-dwellers, such as fishes, lobsters and seals, swallow plastic waste in large pieces.

naturally in rocks at the Earth's surface but arrives from outer space. Meteorites, which are essentially pieces of rock from space, hit the Earth all the time, and these bring tiny quantities of iridium. Most of these meteorites are never noticed, but occasionally one hits a car or a house and the startled inhabitants may take it to their local museum or to the press.

Larger meteorites hit the Earth much less often, but after such strikes, a crater is formed as the meteorite drills deep into the Earth's surface. Generally, the meteorite itself vaporizes and the contained iridium is part of the dust that can be thrown very high in the atmosphere and may travel some distance laterally before it eventually falls back to Earth, possibly during a heavy rain shower. There may be only tiny amounts of iridium in this dust, or clay layer after it settles, a few parts per billion, but it can still be detected. Indeed, at the rock level marking the end of the Cretaceous, a so-called iridium spike or anomaly has been identified at over two hundred localities worldwide on both land and sea. This seems to be an amazingly reliable marker of the massive impact that brought an end to the dinosaurs. DePalma has identified the iridium layer at Tanis just above the fossil layer, so the Tanis fossils really do date from the very end of the Cretaceous. But were they really killed on the last day ... and in late spring?

We will come to that shortly, but first it is important to establish what is known about the end-Cretaceous asteroid impact.

ASTEROID STRIKE: THE EVIDENCE

The meteorite that hit the Earth 66 million years ago was large enough at 10 km (6 miles) across to be classified as an asteroid or minor planet (see Plate XVIII). However it is referred to, it was a large rock and it caused devastation; it seems to have been the only time that such an asteroid impact has caused a mass extinction. The idea was first suggested in 1980, and the evidence for the asteroid has grown since then, with key discoveries in 1990 and 2020, although not all geologists and palaeontologists welcomed the suggestion at first, perhaps as a hangover from the 1950s and 1960s when geologists denied the Earth had ever been hit by large meteorites.

Meteorite impacts were not part of the geology curriculum I studied in the 1970s. In fact, we were warned that such ideas were dangerous or crackpot, and there was little about extraterrestrial geology in the textbooks. Geologists through the earlier years of the twentieth century had even tried to explain obvious craters as volcanic collapse structures. For example, Meteor Crater in Arizona (see Plate XIX), which not only looks like a crater but is even called a crater on the maps, was still debated in the 1940s. Some quoted the views of Grover Karl Gilbert, chief geologist of the United States Geological Survey (USGS) who, in 1891, said the crater had been produced by a volcanic steam explosion. Admittedly, by 1950, most geologists did accept that Meteor Crater had been formed by the impact of a meteorite, but it took the work of two USGS scientists to prove it.

Edward C. T. Chao (1919–2008) and Eugene M. Shoemaker (1928–1997) were both petrologists, experts on the chemical composition and microscopic examination of rocks. They wandered around the floor of the crater and picked up samples of melt rocks (rocks that had been somehow altered by high heat or pressure). Under the microscope, Chao and Shoemaker identified two unusual forms of silica called coesite and stishovite. These were known to occur naturally under conditions of extremely high pressure. Coesite had been synthesized in the laboratory in 1953 (and stishovite was synthesized in 1961); the laboratory syntheses showed that the pressures required were something like ten thousand times atmospheric pressure. Such pressures can arise in nature only from meteorite impact, not from volcanic explosion.

Despite this demonstration, and later work by Chao, Shoemaker and others, it took decades for the evidence of meteorite craters on Earth to be accepted. After 1972, images from the Landsat satellite provided increasing evidence of craters – what had seemed like strange combinations of lumps and hollows in remote places suddenly emerged in the photographs as obvious craters, and by 2015, the count of larger craters (those over 6 km/3¾ miles in diameter) on Earth was determined to be 128. This is a number that would have amazed the early geologists, but it isn't so unexpected given that Mars, the Moon and every other planet is covered with craters. Those on Earth are harder to see because they have been covered by younger rocks or eroded away, or they hide behind vegetation and human development.

So, when in 1980 the Nobel-winning physicist Luis Alvarez (1911–1988) and his geologist son Walter Alvarez and colleagues argued that the Earth had been hit by a 10-km (6-mile) asteroid that killed the dinosaurs, most geologists and palaeontologists were incredulous ... and cross. Incredulous because the idea seemed so far-fetched, and cross because they resented being told their business by a physicist (Walter Alvarez tells the story in detail in what must be one of the best book titles ever, T. rex and the Crater of Doom). Despite this resistance, the evidence has piled up since 1980 and asteroid impact as the dinosaur killer is now the most widely accepted model. There were two key pieces of solid evidence that convinced most – the discovery of the crater in 1990, and the rejection of the Deccan Traps as the cause in 2020.

Even without a crater, the reality of the asteroid impact is accepted because of the worldwide iridium spike, which as we have seen is taken to mark the boundary between the Cretaceous and Palaeogene periods. There is no known process other than the impact of a huge asteroid that could have generated this worldwide iridium-bearing layer, which is found in both marine and terrestrial rocks. It confirms that a dust cloud encircled the Earth and the dust (plus iridium) fell with rain, forming an Earth-wrapping blanket a few millimetres thick. In some places, geologists have also identified coesite and stishovite, and shocked quartz (see Plate xx), in the impact layers. Further, abundant tiny glass spherules in rock deposits around the ancient Caribbean Sea, the proto-Caribbean, have been noted.

In fact, through the 1980s, geologists were homing in on the proto-Caribbean as the place to search for the impact area. The boundaries of the sea were some distance inland over the southern United States and Mexico because, as noted earlier, sea levels then were much higher than today. The accumulations of the glass spherules told the geologists they were close to ground zero. Such glass beads are thrown into the air by an impact as the energy of the impact is dissipated, and they show aspects of the chemistry of the rocks the asteroid hit, in this case marine sedimentary rocks like limestones and even salt beds. They are only found within about 3,000 km (1,860 miles) of the impact site because of the trajectory of the glass beads through the air, initially as hot melt and then, cooling and twirling as they fly, a spherical shape.

The crater was identified in 1990 by Canadian geologist Alan Hildebrand, then a graduate student at the University of Arizona, and colleagues from Mexico and the United States. He was looking at old oilwell subsurface data from Pemex, the Mexican oil exploration company, and spotted evidence for a buried crater that centred on the village of Chicxulub in the Yucatán Peninsula of southeastern Mexico. Subsequent geophysical surveys revealed that the Chicxulub crater, as it was quickly termed, was 180 km (110 miles) across, the size Alvarez and colleagues had predicted in 1980. It had been hidden under younger sediments deposited during the 66 million years since the impact. Drilling programmes brought up melt rocks from the core of the crater that yielded radiometric dates of 66.043 million years ago, exactly the age of the Cretaceous–Palaeogene boundary.

Although the evidence for the asteroid strike had now been generally accepted, some questioned whether it or it alone caused the mass extinction – could the impact have killed the dinosaurs? An alternative hypothesis linked the end-Cretaceous mass extinction to the Deccan Traps in western and northwestern India, great thicknesses of basalt lava that erupted at the same time. If we accept

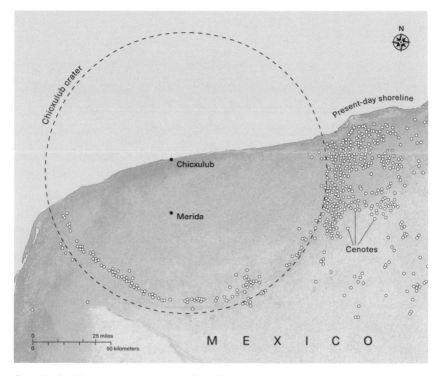

Geophysical image of the huge buried Chicxulub crater, with the northern part beneath the modern sea, and the southern half beneath the land of the Yucatán Peninsula of Mexico.

volcanic models as the driving cause of death in the end-Permian, end-Triassic and other great mass extinctions, why not for the end-Cretaceous catastrophe as well? Doubtless, the Deccan eruptions modified climate and stressed life, but the two were distinguished thanks to a great study in 2020 by palaeontologist Pincelli Hull from Yale University and her collaborators. Using global records of temperature change in the oceans, they showed that the Deccan eruptions peaked before the end of the Cretaceous, generating a sharp temperature rise. The extinctions, however, happened exactly coincident with the asteroid impact, 200,000 years later, when temperatures were driven down. Impacts throw up dust that blocks out the Sun and so causes darkness and cold, and this, argued Hull, was the killer.

BACK TO TANIS

So, what is the evidence that Tanis documents the day the dinosaurs died? First, DePalma and colleagues noted some strange features in the preserved river channels at Tanis, which at the time was located close to the shores of the Western Interior Seaway. It looked as if the river channels had been deeply eroded and that water was flowing in both directions. Something strange had affected these rivers, causing the water to see-saw, flowing rapidly upstream against the normal flow, and then clearing out rapidly to the shallow sea (this is known as a seiche wave, a special kind of side effect of a tsunami or earthquake). They argued that seismic tremors had radiated outwards through the Earth's crust from Chicxulub ground zero as the asteroid struck, reaching Tanis 3,000 km (1,800 miles) away and shuddering the coastline up and down, thus causing the rivers to flow back and forwards through several cycles.

The second piece of evidence is that some of the fossil beds contained glass spherules, just like those found around the impact site, and that they had reached Tanis – possibly near the outer limit of their trajectories – through the air. Some of the glass spherules were located inside the gill cavities of the fishes living in the rivers at the time, and DePalma has suggested they were drawing them in as they swam around in the see-sawing rivers, while the spherules pattered down on the surface of the water. If this is true, then this fixes the timing of deposition of the spherules and the death of the struggling fishes to within an hour or so after the impact.

Other fossils at Tanis include small mammals, a baby pterosaur, the leg of a plant-eating dinosaur called *Thescelosaurus* and a freshwater turtle with a stick passing through its skeleton. Did these die on the day of impact? Possibly. Maybe the turtle was swimming peacefully and then its watery home was upended by the seiche waves and in the torrent of water and debris the poor animal became impaled by the stick.

If we have the killing site, how could DePalma and colleagues say the crisis was in late spring? This is because geology PhD student Melanie During at Uppsala University studied growth rings preserved within the bones of some of the fishes. Sturgeons and paddlefish swam in the Tanis rivers, just as they do today in the rivers and lakes of North America, Europe and Asia. Russian sturgeon are famous for their caviar, and fishermen prize the truly giant 2-m (6-ft)-long specimens they used to catch in Lake Baikal. The Tanis sturgeon were smaller, but their spines show a pattern of stripes at a scale of less than a millimetre, some stripes containing large bone cells, and the narrow ones with very few bone cells. We know from modern fishes that the bone-cell layers represent times of rapid growth when food is plentiful, and the thin layers lacking bone cells indicate winter or times of starvation when growth is slow. During and team studied bones from sturgeon and paddlefish; one key example from a paddlefish showed that it was six years old when it died and stopped growing in late spring (probably May, but possibly late April or June) as indicated

Skull of a paddlefish specimen from the Tanis site. The fish faces right.

by the size of its last growth ring. Its growth was terminated, During and colleagues say, by the asteroid impact.

So much for Tanis, but what do we know about the end-Cretaceous extinctions in general?

DEATH IN THE OCEANS

In the oceans, the main victims of the crisis were the ammonites, belemnites and rudist bivalves. These are all groups of molluscs, and as we have seen, all had been major parts of the marine eco-systems of the Mesozoic. Both ammonites and belemnites were hugely abundant, as attested by their prolific fossils (ammonites comprised hundreds or thousands of species). They formed much of the diet of many fishes and marine reptiles of the Jurassic and Cretaceous, and they sought to escape from their predators by squirting water through their siphons to generate a jet-propelled burst of speed backwards, at the same time ejecting a cloud of black ink to create confusion.

Just as these swimming molluscs disappeared, so too did the great marine reptiles of the Late Cretaceous, notably the plesiosaurs and mosasaurs. In Chapter 7, we saw how this entirely new category of predators entered the oceans in the Early and Middle Triassic, following the end-Permian mass extinction. The key groups were ichthyosaurs and plesiosaurs and they dominated Jurassic seas, as well as, in part, the seas of the Cretaceous. Ichthyosaurs, the dolphin-shaped predatory reptiles, had already died out 28 million years before the end of the Cretaceous (see page 186). Plesiosaurs had divided into two ecological types tra-ditionally called plesiosaurs, which had small heads and long necks, and pliosaurs, which had massive skulls and short necks. The pliosaurs had also died out earlier in the Late Cretaceous, but long-necked elasmosaurids, up to 12 m (40 ft) long and with their snaking necks taking up half that length, survived to the end of the Cretaceous.

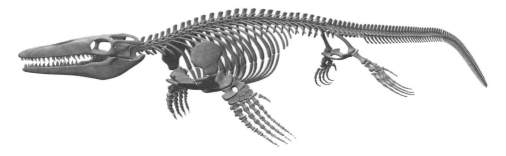

Skeleton of a mosasaur, one of the top predators of Late Cretaceous seas. Mosasaurs were in fact giant lizards, but they died out at the end of the Cretaceous.

The ecological spots in the food webs formerly occupied by ichthyosaurs and pliosaurs were taken over by the mosasaurs, a remarkable group of lizards that became very much adapted for life in the oceans. Mosasaur fossils are found in abundance in the latest Cretaceous limestones of Europe and North America, many of them up to 4 m (13 ft) long. Mosasaurs had torpedo-shaped bodies, a long flat-sided tail with a terminal tail fin used for swimming by sideways sweeps, and four great paddle limbs for steering. Up to the very last day of the Cretaceous, these were the dominant predators, something like killer whales, feeding on fish and smaller marine reptiles, as well as ammonites and belemnites.

Rudist bivalves lived on the seabed, forming great reefs composed sometimes of millions of individuals. Each rudist was about the size and shape of an ice-cream cone, but gnarly and irregular as if the cone had been made by a toddler with a pile of wet clay. Rudists probably lived to a great age, surviving by filtering small particles of food from the shallow waters around them. Other reef-builders such as corals and sponges also suffered losses during the mass extinction, but many species survived. Perhaps the rudists and the other reef-building victims of the extinction were vulnerable to the ocean acidification that happened at the time.

These complex food chains in Late Cretaceous oceans all depended on plankton as their primary food source. Like today, microscopic animals (zooplankton) fed on microscopic plants (phytoplankton), which captured energy by photosynthesis, converting carbon dioxide from the air into carbon to build their bodies and oxygen, which they expired back into the ocean water and the atmosphere. The fossil evidence shows a sharp reduction in plankton populations at the very end of the Cretaceous, and great losses of species of foraminifera and coccolithophores (microscopic organisms with calcite shells).

As a measure of how serious these plankton losses were, let us consider the case of the coccolithophores. These are beautiful fossils when viewed under the microscope (see Plate XXIII). They are ball-shaped in life, each composed of multiple circular plates with radiating spokes. After death, the plates fall apart, forming vast thicknesses of limestone on the seabed. In fact, the White Cliffs of Dover in the UK (see Plate XXII), and similar limestones and chalks of the Late Cretaceous elsewhere, are nearly entirely built of the skeletons of these microscopic algae. So the fact this successful group was nearly wiped out is indicative of the severity of the crisis.

OCEAN ACIDIFICATION

The lack of light for photosynthesis and ocean acidification seem to have been the key factors of the crisis. Lack of light is a well-known consequence of asteroid impact and the blacking out of the Sun. Acidification is something different and has been quite hard to detect. We know that ocean acidification is a concern today, as global temperatures rise; excess carbon dioxide in the atmosphere makes the surface waters of the ocean more acidic, not necessarily sufficient to dissolve the calcareous shells of animals, but certainly enough to make the physiological task of making and maintaining their shells much harder. In more acidified waters, molluscs and corals have to expend more energy than usual to capture calcium ions and build their skeletons. We can recognize acidification episodes in the rock record by the absence of limestones, but there appears to be another independent measure: boron isotopes.

Boron is a naturally occurring element that forms a brown powder and crystallizes as a brittle, dark, shining metal-like material. It occurs in seawater; when this evaporates it can form natural salts, which are mined for use in the glass, ceramic and fibreglass industries. In seawater, boron occurs naturally in two forms, boric acid and the borate ion, and the ratio between these two is a measure of the pH of the water, signifying where it sits on a scale from acid to alkaline.

In 2019, using these ratios, Michael Henehan at the German Research Centre for Geosciences in Potsdam measured the proportions of the two boron isotopes in samples of marine plankton shells through the Late Cretaceous and across the boundary into the Palaeogene. Values indicated reasonably steady boron isotope ratios reflecting a pH of about 7.8, with a sharp spike to 7.5 (marking a build-up of acid) exactly at the Cretaceous–Palaeogene boundary in the immediate aftermath of the asteroid impact. Values then bounced back to 8.0, and recovered to the normal pre-impact value of 7.8 some 100,000 years later. Henehan and colleagues argued that, using an independent method, they had identified that there was a sharp phase of ocean acidification caused by the asteroid impact.

Exactly how the impact fed acid into the oceans is debated, but it is known that the asteroid hit marine limestones and gypsum salt deposits. One effect would have been to send huge volumes of sulphates into the atmosphere that, when mixed with rain, formed sulphuric acid, which would provide a clear acidification effect when it fell into the shallow seas. In addition, huge volumes of carbon dioxide might have been produced by the vaporization of limestones and by wildfires and other consequences of the impact. In the atmosphere, carbon dioxide mixes with rainwater to produce carbonic acid, and this would have further acidified the oceans.

As the acidity levels rose, marine plankton and animals with calcium carbonate shells, including ammonites, belemnites and rudists, died out. In lower pH conditions, they could not build up their skeletons, or their skeletons became too thin for survival. Other sea-dwellers without calcium carbonate skeletons, such as worms, sponges and fishes, were little affected. The same contrast is seen among the planktonic organisms, less familiar to us perhaps, but hugely important in the ocean ecology: as previously noted, the foraminifera and coccolithophores with calcareous shells suffered substantial extinctions, whereas forms with silica skeletons, the dinoflagellates and radiolarians, were not affected.

Because of the extensive loss of species among plankton with calcium carbonate shells, the ocean acidification event detected by Henehan and colleagues resulted in a collapse of the carbon cycling system in the oceans. In the first few tens of thousands of years after the extinction about 90% of foraminifera and coccolithophores disappeared, and photosynthesis and productivity in the ocean fell to about half what it was before the impact event.

Subsequently, even as plankton populations and productivity began to recover, because the new pioneer plankton species that repopulated the oceans after the extinction were small and lightly calcified, the carbon cycle (the system by which the essential element carbon cycles through the bodies of organisms, into the rocks on the seabed and back into the food chain; see page 54) took much longer to return to pre-impact conditions. This is because plankton with heavy shells play a crucial role in clumping decaying plankton remains together and helping to transport them quickly towards the sea floor. When these heavier plankton were lost, organic matter remained in the water column for longer; this favoured the continued dominance of small and lightly calcified plankton, which in turn reinforced this 'new normal' in the aftermath of the extinction. The carbon cycle was thus disturbed for over a million years after the impact.

THINKING ABOUT MASS EXTINCTIONS

The end-Cretaceous mass extinction has been the most-studied crisis of all, generating probably more than a thousand scientific papers each year. This reflects the ready availability of data and, of course, the subject's fascination: anything that includes dinosaurs and meteorite impacts is a sure winner (see Plate XXI). But the event 66 million years ago was a one-off, something quite unusual, and it might be argued we have less to learn from it than from earlier volcano-driven hyperthermal mass extinctions.

There was a time, though, when scientists thought that the end-Cretaceous event was typical of a wider class of phenomena. Soon after the publication of the Alvarez impact theory in 1980, there were flurries of publications by astronomers and others, seeking to identify an extraterrestrial cause, even one that might repeat itself. For a time, people talked of a cyclicity of impacts on Earth, and this idea still has adherents. The suggestion is that every 26 million years a shower of meteorites is released from the outer fringes of the solar system by a long-term cycle of solar system wobble, or even by the influence of a second sun, the so-called death star Nemesis. This would be scary: if we knew there had been an asteroid-driven mass extinction 66 million years ago, then another 40 million years ago, then 14 million years ago, we could predict when the next impact of great magnitude would happen. How would we prepare – should we purchase our hard hats, build underground bunkers, lay in food supplies? Would humans even be around on Earth 12 million years in the future? The evidence for such huge impacts following a regular timing is non-existent, however. Physicists continue to play with tables of extinction totals and find interesting patterns, but the pursuit has become a kind of numerology, seeking pattern where none exists, linking our fates to mystical formulae. Although the next asteroid could arrive any day, we can't predict when.

Do 10-km (6-mile) asteroid impacts always cause mass extinction? The answer appears to be no. In fact, there have been several asteroid strikes on Earth of similar size to the Chicxulub crater, but these did not lead to extinction. It is all dependent on the rocks the asteroid hits. Most of the other giant craters are over inert igneous or metamorphic rocks that, when struck and vaporized, did not release any particularly toxic elements. It was just chance that the Chicxulub impactor hit limestones and sulphur-rich salts, resulting in huge volumes of carbon and sulphur entering the atmosphere and raining acid over the oceans and lands.

This end-Cretaceous catastrophe was the last mass extinction. The mammals – us – were the beneficiaries. The dinosaurs had gone, and despite some speculative promises, it seems unlikely dinosaurs will ever come back to life. What did the world look like in the Palaeocene when the after-effects of the impact had died away? How did life recover and how did the new ecosystems relate to modern ecosystems? We shall explore this in the next chapter.

CHAPTER 13

Recovery and the Building of Modern Ecosystems

PROFESSOR SIMPSON IS PUZZLED

It was 1937 and one of the greatest experts on fossil mammals, American palaeontologist George Gaylord Simpson (1902–1984), was bewildered. He had just published a detailed account of fossils collected from the Fort Union Formation of the Crazy Mountains of central Montana in the USA. What puzzled him was the bone-headedness of the proprietors of the newspapers throughout North America.

Simpson was employed at the American Museum of Natural History (AMNH) in New York and was a restless, enquiring individual, keen to understand and document everything he could about mammal evolution. He once commented, after studying all the mammals from the Jurassic and Cretaceous that were known up to 1940, that they would all fit into his hat (if the museum curators would allow such rough treatment of these rare, tiny fossils). The point was clear – before the end-Cretaceous mass extinction, mammals weren't significant, but after the crisis they were everywhere. The AMNH had been set up as a privately funded museum in 1869, and from the start it took a different approach from the other, public, museums. Professor Henry Fairfield Osborn, president of the institution from 1908 to 1933, decided that it should be a great educational attraction for the public. Osborn was a powerful supporter of using colour reconstructions of dinosaurs and early mammals, whereas curators in university and learned museums were often much more conservative.

Simpson supported this educational function. On the face of it, though, having carefully worked through collections made by other researchers since 1901, his 1937 monograph on the Palaeocene mammals of the Fort Union Formation was a very dry academic work (287 pages describing 57 species of mammals, 37 of which were new to science). From our standpoint today, however, we see that he was describing a remarkably complete and well-documented assemblage of some of the first mammals of the Palaeocene epoch, the time immediately after the end-Cretaceous mass extinction, which was a first important step in helping us understand how life recovered after the crisis.

Simpson knew that there had been a considerable change in ecosystems, with the demise of the dinosaurs and rebuilding of terrestrial ecosystems focused around mammals. He was also a great evolutionary thinker, one of the big names in the 1930s to 1950s who were responsible for the so-called Modern Synthesis. This was when modern evolutionary theory was established, based firmly on Darwin and incorporating the new laboratory-based science of genetics and, after 1953 and the discovery of the structure of DNA, the new science of molecular biology. Simpson, as one of the few palaeontologists with a clear vision of the role of fossils in modern evolution, squashed many of the bizarre theories that had emerged from his predecessors, including his boss at the museum.

As for his bewilderment about the press coverage of his 1937 monograph, Simpson later wrote a delightful short article in which he tracked how an AMNH press summary about the work had been passed from paper to paper, from journalist to journalist, becoming modified along the way. He noted, perhaps rather priggishly, that, 'The [press] release was submitted to me for approval and was issued only after revision seemed to leave no possible false impression. It was a rigidly correct and yet easily comprehensible summary carefully avoiding any sensational claims or mis-statements.' The newspapers had focused largely on the fact that Simpson had described some early primates, distant relatives of modern monkeys, apes and humans, from the Fort Union fauna. The AMNH clipping service collected press reports from ninety-three newspapers; although Simpson felt some of these were good, the majority garbled his message: 'tiny rat is latest ancestor of man'; 'four-inch tree animal seen as man's ancestor'; 'mice of 70,000,000 years ago outlast the dinosaurs'. The good professor despaired.

The time from 66 million years ago to the present day is called the Cenozoic, and it is divided neatly into two, the first half running up to 34 million years ago, when climates were generally warm or very warm, and the second half being distinctly chilly – we will look at the warm post-dinosaur period in this chapter, and the world of cold in the next. One thing we will discover in disentangling events is that today's scientists can offer levels of precision that Simpson would have loved if he had lived long enough to see them.

WARM, WET MONTANA

Climates were warm and wet at the time of the Fort Union mammals. Indeed, climates were as warm and wet as they had been at the end of the Cretaceous when dinosaurs and mammals wandered the banks of coastal rivers, as seen at Tanis (see Chapter 12). Although, as we will see, the end-Cretaceous asteroid impact caused a sharp cooling episode, a few million years after the crisis the landscapes had recovered and the Fort Union mammals experienced warm-climate rainforests of broadleaved angiosperm trees. Indeed, one of the major effects of the end-Cretaceous crisis had been to release much of the evolutionary potential of the angiosperms (see Chapter 11) and enable them to diversify hugely and create a new world, especially in the tropics.

Entering these Fort Union forests, we can see that they buzz with life. Insects flit from flower to flower, and lizards and snakes on the ground and in the trees feed on these insects and on the small mammals that scurry around. In a clearing, 50-cm (20-in.)-long sunfishes move lazily in a warm pond, accompanied by some frogs, salamanders, small turtles and crocodiles. Around the edges, goose-like wading birds delve for molluscs in the mud. All these animals have survived through the end-Cretaceous mass extinction.

A remarkable mammal suddenly rushes across the clearing. It looks like a small person, running on its hind legs, body leaning forward and arms clutching a juicy, struggling cockroach. It has a long snout, small sharp teeth in its jaws and large eyes that peer alertly from right to left. This is *Prodiacodon*, a leptictid, one of the unusual mammal groups of the Palaeocene that did not give rise to any of the modern mammals. With its long, slender tail, it can scurry along on its hind legs for considerable distances, using its arms to carry objects or help swing itself up into a tree. It has reason to be fearful. It is being pursued by a *Didymictis*, a slender cat-like predator called a viverravid. Typical of the Palaeocene worldwide, viverravids are successful as medium-sized predators and distantly related to modern carnivores such as cats, dogs and bears. The *Didymictis* lopes after the *Prodiacodon*, freezing as its prey looks behind, then leaping in for the kill. The *Prodiacodon* drops its burden and shoots up a tree. The *Didymictis* looks at the mangled cockroach, still wriggling its legs, sniffs and walks off, then comes back and lazily picks it up. It chomps, crushing the cockroach exoskeleton and spits it out with a grunt of disgust.

In the tree above, a pair of primates, *Paromomys*, watch the scene without moving. Distant relatives of monkeys and humans, they

Skeleton of the small hopping leptictid *Leptictidium* from the Eocene of Germany, close relative of *Prodiacodon* from the Fort Union beds.

look more like squirrels, with their short snouts, strong limbs and long fluffy tails. They have large heads and large, intelligent-looking eyes. The eyes face generally forward, and each eye's range of vision overlaps so they can see a single scene in three dimensions, so-called binocular vision. This is important for these tree-dwellers because they have to be able to judge distances accurately before they leap from one tree to another. Earlier primates with dog-like snouts had to flick their heads from side to side to interpret the scene; they did not dare leap towards a distant branch because they might misjudge the distance and end in a heap of broken bones on the ground far below.

In a clearing nearby are two larger mammals. *Tetraclaenodon* is a cat-sized, horse-shaped mammal that feeds on tree leaves. It is classified as a phenacodont, a somewhat mysterious group that includes the relatives of modern horses and rhinos. Slightly larger is the sheep-sized *Pantolambda*, which pads about on the forest floor, stepping slowly with widely spread five-toed feet. It is a pantodont, another early mammal group without direct living relatives, which has a large head and deep jaws, adapted for chomping tough vegetation.

Both herbivores are reminders of the diversity of mammalian experimentation in the Palaeocene, with some belonging to modern groups, others to groups that were successful for a while, but then disappeared. The fifty-seven species of mammals in the Fort Union Formation identified by Simpson ranged in size from tiny shrew-like creatures that would sit neatly on the palm of your hand to the sheep-sized *Pantolambda*. This large diversity of mammals is all the more remarkable when we recall that only five million years had elapsed since the end-Cretaceous asteroid wrought huge destruction of life on land. This event is often credited with having created the foundations of the modern world: dinosaurs out, mammals in. Is this true and is it possible to document anywhere exactly what happened between the collapse of the dinosaur-dominated ecosystems such as Tanis and the new, mammal-dominated ecosystems such as at Fort Union?

DINOSAURS GIVE WAY TO MAMMALS
IN THE DENVER BASIN

American palaeontologist Tyler Lyson of the Denver Museum of Nature & Science led a remarkable study published in 2019. He and his collaborators focused their attention on a 27-sq. km (10-sq. mile) rocky outcrop in the Denver Basin, Colorado, called Corral Bluffs. There they found fossil mammals and other vertebrates at 299 locations, and fossil plants at 65. Most importantly, they were able to show that the 240-m (790-ft)-thick rock section spanned the Cretaceous–Palaeocene boundary, comprising mudstones and sandstones deposited by rivers that had flowed widely over the landscape, picking up dead animals from the ecosystems on the river banks, in ponds and in the surrounding landscapes.

The geological age dating is remarkably good. Lyson and colleagues were able to use a combination of magnetostratigraphy and exact radioisotopic ages to pin down ages to within tens of thousands of years. Magnetostratigraphy is based on the fact that the Earth's magnetic north pole flips to the south from time to time, and it flipped three times during this particular interval. The times of magnetic reversal are recorded by magnetization of iron minerals in the rocks and form reliable markers of a particular narrow interval of time. In addition, radioisotopic ages are measured from minerals in volcanic ash beds that record the instant of cooling of the molten minerals; the balance of proportions of different radioactive elements along a radioactive decay sequence can give a measure of the age in millions of years.

Lyson's first observation was that the immediate post-impact cold snap lasted for a very short time, geologically speaking, about 100,000 years. Temperatures fell by 5°C (9°F) during this time, as estimated from leaf data, and this corresponds to findings elsewhere both on land and in the sea. This cold episode following the asteroid impact was not the short cooling caused by the dust of the impact, but a longer-lasting downturn in temperature. In Chapter 12, we saw that this cold spell was associated with catastrophic acid rain on land, which killed plants and life in general, and also caused acidification of shallow oceans, which killed much of the life of the seabed.

The Corral Bluffs sections show that following this the Palaeocene began with a warming episode of 100,000 years during which temperatures rose by 5°C (9°F), recovering to pre-impact levels. Temperatures rose by a further 3°C (5.4°F) 300,000 and then again 700,000 years after the crisis, ending up overall about 10°C (18°F) warmer than immediately after the impact. These higher temperatures provided great opportunities for life to diversify.

At the end of the Cretaceous, the diversity of tree species had halved, and many ecological types disappeared completely. However, by the time of the second warming episode, 300,000 years into the Palaeocene, flowering plant diversity had recovered to nearly double pre-extinction levels. New angiosperm groups had emerged, including the oldest example of Fabaceae (sometimes called Leguminosae; the bean family), now represented by

20,000 species. As we will see, this rapid recovery of warm, humid forests was key to the building of the modern world.

Mammals showed a rapid recovery, too, and the fossil evidence is outstanding, with exceptionally well-preserved, three-dimensional skulls and other bones preserved in rocky concretions. Through the Corral Bluffs rock sequence, Lyson and colleagues noted that the larger mammals died out at the end of the Cretaceous and only smaller forms survived. But, again, within 100,000 years, the survivors had become larger, equalling pre-extinction sizes of up to 10 kg (22 lb). Then, 700,000 years after the crisis, the mammals were up to 40 kg (90 lb), the size of a German Shepherd dog. At this same time, mammals were diversifying, from being primarily insect-eaters just after the asteroid impact to a mix of insect-eaters and plant-eaters. The larger, plant-eating mammals benefited from the diversification of angiosperms, which provided delicious new foods such as beans, and there is also evidence for migration and mixing of faunas and floras over long distances.

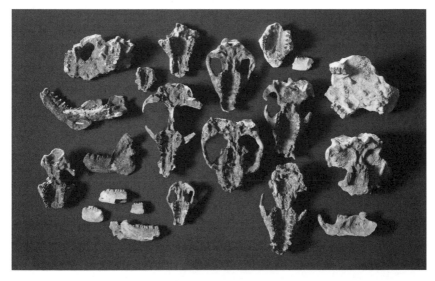

A selection of the skulls and jaws of early Palaeocene mammals from the Corral Bluffs sections in Colorado, some of the first to experience life without the dinosaurs breathing down their necks.

Compared to the time life takes to recover after hyperthermal events of any size, the Palaeocene recovery was fast. This is probably because asteroid impacts can cause short-lasting, devastating effects such as global cooling and acid rain, but no long-lasting consequences, so the Earth recovers quickly. Hyperthermal events, on the other hand, sometimes show successions of 'after shocks' such as further eruptions and warming pulses, or the carbon and other cycles are so out of balance, it takes longer for them to recover. The Corral Bluffs sections show a short-lived cold episode of ferns, conifers and small insectivorous mammals, before life bounced back remarkably fast, within only 100,000 years. By the time of Simpson's Fort Union faunas, five million years later, some terrestrial ecosystems were already about half as rich as any modern tropical rainforest.

GIANT BEANS OF GREAT WORTH

Beans are seeds, and these new plant groups evolved and rapidly became a thousand times larger, all in the Palaeocene. Indeed, the average size of angiosperm seeds increased from 1 cu. mm in the Late Cretaceous, the size of a coriander seed, to 10 cu. cm (½ cu. in.), the size of a plum, in the Palaeocene. Why this amazing change? When palaeobotanist Fabiany Herrera of the Field Museum in Chicago reported the world's oldest giant fossil bean from the Palaeocene of Colombia in South America, he had a specimen of much greater value to a palaeontologist than the magic beans of Jack and the Beanstalk. The large pod contained six juicy specimens as large as broad beans and was quite different from anything that had existed five million years earlier, in the latest Cretaceous. The enormous change in tropical rainforests had happened, and the nutritious beans of the newly abundant Fabaceae were an important part of the Palaeocene boom in tropical rainforests, stimulating the evolution both of plants and animals.

As discussed in Chapter 11, the beans were food for both the developing plant seeds and for animals. Angiosperms had evolved

Fossil beans from the Cerrejón Formation of Colombia, some of the first large angiosperm seeds ever, marking the origin of modern-style tropical rainforests.

to exploit animals as carriers. Tiny Cretaceous seeds simply exploded from their capsules or pods and fell close to the parent plants. Legumes with nutritious beans had seeds that were an attractive food for animals large and small, who would then carry them widely and eventually deposit them in their dung, providing the seed with its own dose of compost. This mode of seed dispersal was one of many coevolutionary relationships between flowering plants and animals, and it may be that bean-eating animals in the Palaeocene were becoming picky and actual bean size mattered in attracting the bean-eaters.

From their beginnings in the Cretaceous, angiosperms had evolved in concert with various insect groups to distribute their pollen to enable cross-fertilization of the developing seeds. Continuing the Cretaceous theme, the Palaeocene coevolution of seeds and seed-dispersers was another strong link between plants and animals, and this time the animals were mainly mammals and birds. The great ecological success of angiosperms was enabled by these measures to improve cross-fertilization and seed dispersal.

The plant–animal coevolutionary relationships themselves drove specialization and species division. Perhaps, in the Palaeocene, pollen and seed spreading was done by numerous species of insects and mammals that could access these food sources from a single angiosperm species, but it would then become inevitable that specialization would occur, as one insect or mammal species became ever better adapted to focus on securing food (and thereby scattering pollen or seeds) from a particular flower species. This process of specialization drove further species splitting. Tropical rainforests have very high biodiversity for these two reasons: more energy and food as a result of greater levels of sunlight capture by their leaves (see page 187), but also high levels of species-splitting thanks to the sharpening of plant–animal coevolution.

What part, if any, did the loss of dinosaurs play in the origin of these astonishing modern tropical rainforests?

DINOSAURS VS. RAINFORESTS

Today, much of terrestrial biodiversity is to be found in the rainforests, particularly the huge rainforests of Central and South America. There are thousands of tropical hardwood tree species, each with its own remarkable array of specialized insects, spiders, mites and centipedes. These are preyed on by a dazzling panoply of frogs, lizards, snakes, mammals and birds. Much of the richness of modern life had its origins in the Angiosperm Terrestrial Revolution that we explored in Chapter 11, but the final phase had to await the demise of the dinosaurs. The key themes of the second step in this revolution were the dramatic increase in seed size as a marker of the rainforest boom as discussed above, and how dinosaurs had previously suppressed these changes.

Palaeobotanist Mónica Carvalho, then of the Smithsonian Tropical Research Institute in Panama and now at the University of Michigan, and collaborators, showed in 2021 that tropical rainforests boomed as a result of the end-Cretaceous mass extinction.

Based on their studies in Colombia, the site of the giant Palaeocene beans, they found that plants declined to nearly half their pre-extinction diversity and took about six million years to recover. The upward swing in biodiversity did not stop at the levels of the latest Cretaceous, however, but continued rising; equivalent biodiversities today would be ten to a hundred times higher than they had been in the Cretaceous. The new tropical rainforests had a multiple-layered structure, with different species occupying different vertical levels, from the ground to the tops of the tallest trees. Remarkably, the same plant families we see today were already there, and in similar proportions. In addition, the tree canopy was closed, as it is today, without many open spaces. When a tropical tree falls, its neighbours quickly plug the gap by growing into the sunlight.

Carvalho and colleagues stated that 'the end-Cretaceous shaped modern Neotropical rainforests', and that the extinction crisis did this in three ways. First, the conifers that had been important in the Late Cretaceous floras of the Americas, such as monkey puzzle trees (Araucariaceae), died out or became greatly reduced in diversity and so gave the surviving angiosperms the chance to be the tallest, thereby forming the architecture and ecological proportions of the new forest canopy. Second, they argued that the soils in many of these Late Cretaceous forests were poorly fertile, but the ash fallout from the Chicxulub impact spread phosphates and other chemicals that enriched the soil; this favoured the growth of angiosperms at the cost of conifers, which live in poorer, infertile soils. The scale of this asteroid ash fertilization effect, though, is uncertain. The new angiosperms, such as Fabaceae with their beans, further fertilized the soils and made them richer. Third, the dinosaurs played a role: in their heyday, they trampled around knocking over bushes and trees, as elephants do today, and so kept terrestrial landscapes open and felled; they also ate any trees before they got too large. With dino-saurs gone, the forests filled in, closing the gaps, and took over the landscape.

THE PETM: A DIVERSITY-ENHANCING HYPERTHERMAL?

Through the 66 million years of the Cenozoic Era, there were a number of major climate changes, some associated with small-scale hyperthermal events. Hyperthermal events were generally killers as we have seen, but the last in a long series inexplicably didn't cause much lasting extinction and actually seems to have enhanced biodiversity. This was the Palaeocene–Eocene Thermal Maximum (PETM), a time of sharp changes in climates at the end of the Palaeocene, 56 million years ago. Temperatures rose by as much as 6°C (11°F), and there were extinctions among deep-sea plankton, deep-sea-dwellers and especially the terrestrial mammals. However, it appears that plankton made a fast recovery, if hit much at all, and mammals were stimulated to increase in diversity.

This was perhaps the last substantial phase of global warming and ocean acidification to have been driven by volcanic eruptions. This time, the offending eruptions were happening in the North Atlantic Igneous Province (NAIP), centred on Iceland, which sent out huge volumes of lava and volcanic gases. Its effects are especially well studied in Northern Ireland, where the lavas include those of the Giant's Causeway, and in the west coast of Scotland, where its lavas form the famous basalt columns seen in Fingal's Cave on the island of Staffa.

During the PETM, sea levels rose because of thermal expansion – where water occupies more space as it warms up. Warming of deep waters in the ocean and plankton blooms led to perturbations in water circulation patterns and some ocean floors became anoxic, leading to the death of seabed animals. Ocean acidification killed corals and other animals that had calcareous skeletons but the levels of extinction among plankton and creatures that swam were much lower than might have been expected.

On land, the PETM warming led to drier climates as shown by detailed studies in the Bighorn Basin of Wyoming. Mammal faunas

Fingal's Cave shows columns of basalt lava that were erupted 56 million years ago across a wide area of the North Atlantic.

in North America changed markedly, mainly because of invasions of new mammal groups from Asia and Europe. The Atlantic Ocean at the time was much narrower than today, and in any case, warmer climates enabled animals to migrate across the previously too cold Bering Strait from Russia to North America, and through Greenland from Europe to North America. But, although these movements of mammals substantially changed the faunas, it seems most of the invaders were accommodated into the ecosystems without extensive extinctions among the established groups. This was because the new habitats offered so many ecological options that the invaders were able to establish themselves without much conflict.

The PETM was in some way a non-extinction event. Global warming and climate change on land actually led to increased biodiversity, especially among mammals. Turnovers of species in the sea certainly happened, but many groups recovered quite quickly and rose to higher diversity as temperatures continued to rise in the Eocene. There are still questions, however, about how the hyperthermal mechanism of this event operated and how the eruptions caused such a level of warming.

THE METHANE GUN: TRUE OR FALSE?

Geologists realized early on that the NAIP eruptions would not have provided enough carbon dioxide to drive the 6°C (11°F) of global warming, and another source was needed. In this case, and perhaps in other hyperthermals (see Chapters 6 and 10), this was methane. In the late Palaeocene, great volumes of methane gas had been locked into frozen reservoirs in the deep oceans around the margins of major continents. These frozen water structures are called methane clathrates, which exist under high pressure at depths of 500 m (1,600 ft) or more. If the oceans warm a little, the clathrates begin to melt and rise, and as they do, they can expand 160-fold, releasing a huge amount of methane gas into the atmosphere. Methane is a powerful greenhouse gas that causes considerable warming when it enters the atmosphere (see page 170).

There is a risk of positive reinforcement in the methane clathrate model. If methane is released at depth and explodes upwards to enter the atmosphere, air temperatures rise. This further warms the ocean and could cause more release of methane from depth, and further warming. This has been called the 'runaway greenhouse effect' or the 'methane gun', highlighting the fact that deep-ocean methane release could be unstoppable after the process is started. Far from the usual negative feedback controls in nature that generally slow such processes down or reverse them, here the feedbacks are positive: warming leading to ever more warming.

On the discovery of this methane phenomenon, some climate commentators have suggested that what happened during the PETM is also happening now, and that once methane release starts, it will never stop until the Earth has burnt itself up. But did the Earth face such an out-of-control system during the PETM? And is this what we can expect in the future? The answer to both questions is probably no. Carolyn Ruppel, an expert on deep oceans and methane who works at the United States Geological Survey, notes that the volumes of buried methane clathrates may be lower than sometimes

221

estimated, and that releases of methane from depth don't all hurtle up to the surface at the speed of an express train. Much of the methane gas is absorbed into the water as it passes upwards and so does not reach the surface. Further, the reactions that release methane from its icy prison absorb heat from the surrounding seawater, which dampens down the reaction rate. As Dr Ruppel notes, 'This is a problem when we try to produce methane from hydrates – it keeps shutting itself down.'

The story of the Palaeogene has been one of warming, but everything changed at the end of the Eocene. Climates became cooler, and cool climates stimulated the expansion of grasslands on most continents, which provoked a dramatic switch in mammalian evolution from leaf-eating to grass-eating. This might also have flushed a particular lineage of primates out of the forests and onto the open plains of Africa and been the trigger for human evolution. We explore these important episodes of extinction and diversification in the last 34 million years in the next chapter.

PART 6

The Oligocene to the Present Day

34 million years ago–

CHAPTER 14

Cooling Earth

THE FORT TERNAN SAVANNAH

Fort Ternan has been the focus of palaeontological attention since the early twentieth century. Today, it is a tiny, spread-out village in the west of Kenya, 60 km (40 miles) from the shores of Lake Victoria and near the international border with Uganda. It is a trading point for the region, equipped with schools, churches and a district hospital. The village is surrounded by rich farmland where the rainfall is sufficient for farmers to grow all the staple food crops: maize, beans, sweet potato, sorghum and cassava.

If we zoom back to 14 million years ago, the maize and sweet corn fields are lush grasslands: this middle Miocene world is very different from the Palaeocene, when most terrestrial ecosystems were forests of some kind. Climates are cooler, grasslands have spread widely and new mammals are on the scene. Looking out over the Fort Ternan plain, we see a rich savannah of tall and short grasses, with bushes and small trees scattered here and there, like the Queen Elizabeth National Park in Uganda today, where Uganda kob, waterbuck and topi graze, duikers and bushbuck browse on the leaves from bushes, elephants, buffalo, hippopotamus, warthogs and Nile crocodiles are found in and around the small lakes, and the main predators are lions, leopards and hyaenas.

Under the blazing Sun, there is a shimmering haze all around, and unseen crickets chitter and chirp, making a continuous wall of sound. In the foreground are modern-looking antelope, including species of spiral-horned *Oioceros* and straight-horned *Kipsigicerus*. They are feeding in mixed herds of dozens or hundreds of individuals, heads down, snipping grass and wrapping their tongues round the tufts before passing the stalks back in their mouths for a thorough chewing. These ruminants, related to modern antelope and cattle, are capable of processing grass in multiple cycles, passing the food into the first stomach where it resides for a while, before being regurgitated for further chewing. Like cattle, they chew the cud. Ruminants have four-chambered stomachs and the chewed plant food passes through all four, ensuring that these animals secure maximum nutrition from the grass. Another necessary adaptation is needed to feed on grass, which contains silica in its structure (itself presumably an evolutionary defence against herbivore browsing) and, possibly more importantly, dust and grit from the ground. Any animal that does not have specially adapted teeth would soon wear them down – ruminants, such as cattle, giraffes, antelopes and deer, and their wider relatives, including camels and pigs, have high-crowned teeth to cope with this constant rasping of silica and grit.

Small groups of another ruminant, *Samotherium*, stalk between the trees in the distance, at the edge of the plain. Their slender bodies and long limbs are as large as those of a modern giraffe, but the neck is half the length, about 1 m (3 ft) long. Modern giraffes can be 6 m (20 ft) tall, while *Samotherium* is 4 m (13 ft) tall, but it makes up for its smaller stature with some large, straight horns on top of its head. It feeds on leaves from bushes near the ground, but mainly concentrates on stretching up to feed on trees that are out of the reach of the other herbivores. Africa and Europe are not at this time separated by the Sahara Desert, so the lush grasslands are continuous to the shores of the Mediterranean and these early giraffes and antelope roam throughout Africa and north as far as Greece and Italy.

The huge predatory hyaenodont *Megistotherium*, twice the size of a lion and the terror of the African savanna 14 million years ago.

Suddenly, the antelope raise their heads sharply and shuffle. The *Samotherium* melt into the trees. The long grass rustles in the distance, but the threat is not at first visible. Some antelope on the edges of the herds begin to scatter. Then, an enormous head emerges from the grass, staring fixedly at a young antelope. This is something unfamiliar, something prehistoric, completely unlike anything alive today. Its head is 66 cm (2 ft 2 in.) long, with a long snout armed with whiskers and penetrating eyes. The jaws are lined with enormous teeth that justify its name – *Megistotherium osteothlastes*, meaning 'very large beast, bone crusher'. The teeth are yellow and brown, and the beast drools gobbets of slobber from the edges of its mouth. This is a huge predatory hyaenodont, a survivor from the Palaeocene and not closely related to any modern predators. It advances further, padding softly, and brings its entire half-tonne body into view; it is twice the size of a modern male lion, and a threat to every animal on the plain. The huge predator continues walking slowly towards the antelope, but they are nimble and skip aside from its path. It breaks into a trot, eyes fixed on the young animal it had spotted, but the antelope soon escapes. The hyaenodont lacks the speed and agility to catch any but wounded animals; roars of frustration rumble from deep in its belly and set the plain alive with their reverberations.

Some wild pigs run by, followed by a short-legged rhinoceros, Paradiceros, which is smaller than its modern relatives. They are heading for the water hole, where crocodiles, turtles and white wading birds pick around for food. Towering over them are two very extraordinary-looking elephant relatives. One, Platybelodon, is 3 m (10 ft) tall, the size of a present-day female African elephant, but it bears tusks above and below. Its lower tusks are in fact enormously expanded, shovel-like incisor teeth on its lower jaw. It is swirling its long goofy lower jaw in the water, stirring up the mud, but generally uses the upper tusks and lower scooping plate to grasp tree branches and bundle them together before pushing them into its mouth with its short trunk. Beside it are a couple of Deinotherium, larger than the Platybelodon and lacking upper tusks, but with lower incisors bent back under the jaw as tusks, perhaps for de-barking trees.

At first sight, nothing seems to be living in the short, scrubby trees near the water hole, but behind some leaves, two round, intelligent eyes stare out at the elephants. Hiding in the shade is an early ape, Proconsul, about the size of a small chimp and weighing about 18 kg (40 lb). It has a broad snout and large eyes, and its teeth suggest a diet of fruit. The Proconsul now shows its skills, loping along a thick branch on all fours, long arms and legs moving fast. It swings beneath a branch and drops to the ground, scooping up some fallen fruits in its hands and leaps back into the tree to eat them.

That we can picture this scene is made possible by the collections of thousands of specimens at the fossil sites around Fort Ternan. Many of these Miocene mammal species are known from other localities, too, which has enabled detailed study of their taxonomy and function. Experts have focused on, and long debated and disputed, the human-like primates – do we call this beast *Proconsul*, *Kenyapithecus* or *Ekembo*? Is it an advanced monkey or a primitive ape? Could it run, walk, climb or swing along under the tree branches like a modern ape? Attention has also been paid to the palaeoclimates and plants of the time. How was this ecosystem affected by cooling climates, and how does this relate to the modern world?

THE GRANDE COUPURE

The Fort Ternan savannah was part of the new cooler world that began 34 million years ago, at the time of the Grande Coupure. Temperatures worldwide in the Palaeocene and Eocene had been generally warm (see Chapter 13), but there was a step down in temperatures that has never reversed. We have moved from the 'greenhouse world' of the dinosaurs and early mammals to the current 'icehouse world'. This has greatly shaped the modern world, and it is worth trying to pick apart what happened and why. What was this event, and why is it called the Grande Coupure?

The second half of the Cenozoic began with a short global warming episode followed by a permanent drop in temperature marking long-term cooling. This Eocene–Oligocene extinction event, 33.9 million years ago, includes several components, one of which was named the Grande Coupure ('the great break') by Swiss palaeontologist Hans Georg Stehlin (1870–1941), who in 1910 spotted a time of rapid turnover in the mammal faunas of Europe. Stehlin had begun his career as a medical doctor, graduating with a PhD on the embryonic development of ruminant mammals, written in German. Armed with a huge moustache, he had a severe appearance, but apparently debated and discussed all subjects with a merry twinkle in his eye, especially delighting in the subtleties of the French language. In his studies of the Eocene and Oligocene of the Paris Basin, Stehlin had noticed that the European-style mammals of the Eocene were infiltrated by Asian-style mammal groups such as rhinoceroses, giant pigs, new rodents (hamsters and beavers) and new insectivores (hedgehogs). There were also widespread extinctions of some medium to large herbivores.

Causes of this event are complex, but there was certainly a cooling event around the time of the mammalian turnover. Perhaps these lower temperatures were triggered by meteorite impacts, marked both by the Chesapeake Bay crater in North America and the Popigai crater in Russia? The dating of these meteorite impacts suggests, however, that they occurred some time before the cooling. The more likely cause seems to be related to a sharp 40% reduction in the levels of carbon dioxide in the atmosphere, which led to more extensive ice caps at the North and South Poles, which in turn led to further cooling.

The Eocene–Oligocene cooling event marks a turning point in global climates, when the warm phases of the Cretaceous and first half of the Cenozoic came to an end, and temperatures continued declining more or less continuously to the present day. The world changed in profound ways.

ICE SHEETS AND OCEAN CURRENTS

The most dramatic change that began in the Oligocene was the formation of permanent polar ice sheets. Ice had doubtless been present during winter at the Poles before this time, but it melted in summer. Now, the ice stayed put all year; the modern Antarctic ice sheet began to grow at the beginning of the Oligocene and has never stopped growing. This coincided with the formation of the Antarctic Circumpolar Current (ACC), the powerful water movement that swirls around Antarctica, running from west to east. Why west to east? That's the way the Earth spins, and as it spins, the mass of ocean water at the Equator and Poles is dragged in the same direction.

Why did the ACC get going at this time? During the early Cenozoic, Antarctica lay across the South Pole, but the very southern tip of South America was still fused to the Antarctic land mass. In the late Oligocene, about 25 million years ago, the two continents finally split apart, and the Drake Passage opened. Westerly-flowing cold waters that had hit the barrier and shot up the west coast of South America could suddenly surge through and continue all round the world. The ACC has bedevilled mariners since the days of polar explorers Clark Ross, Cook, Scott, Amundsen, Shackleton and all the others: you can go with the flow and sail from South America to southern Africa, from there to Australia, and back to the southern tip of South America. But the reverse, from east to west, can be more problematic, partly because of the current but also because of the prevailing westerly winds.

The northern Greenland ice sheet came much later. There were small ice sheets over the North Pole from 18 million years ago, but the present, large ice sheet began to grow dramatically in extent only five million years ago, and this was the prelude to the great northern Ice Age – the time of the mammoths, woolly rhinos and stocky Neanderthal peoples, to which we will come a little later. This is the world we know today, so two key questions are, how rapidly did global temperatures fall 34 million years ago, and why?

HOW COLD, HOW FAST AND WHY?

Marine records show that the cold phase was initiated by a sudden drop in temperature by 5.4°C (9.7°F). Marine palaeontologist Helen Coxall from the University of Stockholm and colleagues studied oxygen isotope records from foraminifera, microscopic organisms with calcite shells, and in particular those that lived on the seabed. The ratio of the abundant ^{16}O isotope to the rare ^{18}O isotope gives a measure of water temperature. This is because ^{16}O evaporates first into water vapour as water is warmed up, and the remaining liquid water is then enriched in ^{18}O, so the ratio of the two isotopes can be calibrated to indicate temperature. They found that ocean temperatures fell in two steps of about 40,000 years each, during an overall interval of 300,000 years, and this is seen in rock cores taken from the floors of the Southern, Atlantic and Pacific oceans, indicating it was a worldwide change.

Coxall and colleagues also identified a remarkable change in the carbonate compensation depth (CCD) – the level in the deep oceans at which calcium carbonate dissolves under pressure. Above the CCD, animals such as foraminifera, molluscs and corals are happy because their shells remain intact; below the CCD, their shells dissolve and they are most uncomfortable. The CCD today is at a depth of about 4.5 km (15,000 ft); at the point of sharp cooling 34 million years ago, the CCD plunged downwards from 3.5 km (11,500 ft) to the present level, which is a pretty amazing shift of 1 km (3,280 ft) and points to some dramatic change in the oceans. This episode has been connected to the growth of the Antarctic ice sheet, but Coxall and colleagues point out that it occurred world-wide, and so also implies some glaciation at the North Pole, too.

The depth of the CCD depends on the balance between the acidity and alkalinity of the oceans (its pH value). Pure water has a pH of 7 when it is neither acid nor alkaline; values below 7 indicate the build-up of acid and above 7, an increase in alkalinity. Seawater today has a pH of 8.1, so it is slightly alkaline, but as carbon dioxide levels in the atmosphere build up, the values fall towards the acid end of the scale. At the beginning of the Oligocene, the growth of the Antarctic ice sheet removed water from the oceans and turned it into ice. Sea levels fell worldwide, exposing large areas that had been continental shelf. This reduced the production of calcium carbonate by corals and other shallow-water animals, and made the oceans more alkaline as the animals stopped extracting calcium carbonate from the water. Also, as the continental shelf was exposed, great layers of limestone around the coasts were open to erosion, so washing more calcium carbonate into the oceans. Perhaps these two effects drove ocean alkalinity up fast, and so forced the CCD rapidly down.

The same rapid drop of temperature was also identified on land. A 2021 paper led by geochemist Vittoria Lauretano from the University of Bristol and colleagues showed that the chemistry of lignite (soft brown coal) from Australia indicated the same sharp reduction in temperature. The team studied specific lipids (a broad range of fatty and waxy substances found in living things that form membranes and are involved in storing energy) in the coal that indicated original pH and temperature at the time of deposition. They found the same sharp 5°C (9°F) drop as in the oceans. So, the sharp cooling happened on land as well as in the seas, but what was the underlying cause? Lauretano and colleagues tested various possible explanations for the initiation of the cold phase of the last 34 million years by experimenting with Earth-ocean computer simulations. These showed that the main cause was a reduction in the proportion of carbon dioxide in the atmosphere.

But what caused the sharp and permanent drop in carbon dioxide levels? In 2017, climate scientist Geneviève Elsworth of McGill University in Montreal and colleagues tied the change to the opening of the Drake Passage and changes in the circulation of water in the Atlantic. When the Drake Passage opened, the ACC began to swirl round the initiating southern ice sheet and ocean circulations changed dramatically. A short-term warming and rainy phase increased the weathering of rocks worldwide. Weathering includes all the processes that break up and dissolve rocks on the land by rainfall and daily and seasonal temperature changes, and is a key part of the global carbonate-silicate cycle.

The carbonate-silicate cycle links the Earth's crust, the oceans and the atmosphere, and describes how carbon and other elements remain in balance. Carbon dioxide enters the atmosphere primarily through photosynthesis and volcanic eruptions and is returned to the Earth's crust through weathering. This is because silicate rocks, like granites and sandstones, break down by absorbing water and carbon dioxide. The dissolved mixture is washed down rivers into the sea and eventually settles on the ocean floor, while also providing nutrients for marine organisms to build their skeletons and shells. At times of high weathering rates, such as 34 million years ago when sea levels had suddenly lowered, a large amount of carbon dioxide is absorbed by high weathering rates and, in this case, it seems the reduction of carbon dioxide in the atmosphere was permanent, locking the Earth into temperature conditions at least 5°C (9°F) cooler than had been the case in the early Cenozoic.

Why did temperatures not warm up again? This is not clear, but there is something permanent about large polar ice sheets: cold begets cold. The cooler temperatures led to reductions in rainfall, and the lush rainforests in tropical and temperate areas retreated and were replaced by extensive grasslands, as at Fort Ternan and, indeed, today over central parts of Africa, North and South America, Asia and Australia. A classic model of human evolution proposes that this was the time when our ancestors were flushed out of the trees and into the open. Is this true?

WALKING UPRIGHT

The history of anthropology changed forever in the 1920s. That decade marks a watershed in theories and in access to fossils, and the two are linked. With limited fossil evidence, anthropologists before 1930 generally argued for a brain-first model, where our ancestors distinguished themselves from our close relatives, the chimps and gorillas of Africa, by being brainy. This is a nice idea, but untrue. After 1930, with increasing collections of early human fossils from Africa, it became clear that for much of human evolution, our ancestors were not exactly super-intelligent, but they did walk upright.

It's easy to tell whether an ape fossil is from a quadruped or biped. Chimps and gorillas have huge, powerful arms for knuckle walking and occasionally swinging through the trees. They can run upright, but their running is a bit uncontrolled as they stagger along on short, bandy legs. Humans, ancient and modern, quickly adopted long, straight-up-and-down legs, and switched from climbing around in the safety of the trees to walking out in the open and using their arms for carrying things. Every detail of their legs and arms alters, so even fragmentary fossils, like a single toe bone or a scrap of the hip, can immediately reveal posture and locomotion.

As more and more fossils were found from the grasslands of the Miocene and Pliocene of East and South Africa, it became clear that Africa was the cradle of humanity. *Proconsul* from Fort Ternan was one of those precursors of modern-type apes and humans and was definitely a tree-dweller. The search for early human fossils in Africa escalated through the 1920s and 1930s and has continued as a very active field since.

What we know as a result of all this work is that most of the seven million years of human evolution took place in Africa. It was there that the first upright apes stepped out onto the open grasslands and began the long evolutionary divergence from chimps and gorillas, who largely stayed behind in the forests, where they still live today. Multiple species of *Australopithecus* and relatives colonized much of Africa, and one lineage eventually gave rise to our own genus, *Homo*, just over two million years ago.

The first humans to leave Africa were members of the species *Homo erectus*, who spread across the Arabian Peninsula about 1.6 million years ago into Central Asia and across to China and Java, and then became extinct. Various archaic species such as the Neanderthals entered the Middle East and Europe at the time of the Pleistocene Ice Age, and eventually modern humans, *H. sapiens*, evolved in Africa 300,000 years ago, spreading worldwide about 100,000 years ago. All the other dozens of human species died out and humans today, in all their variety and diversity, are members of that single species. These facts seem clear, but the road to understanding has been rocky.

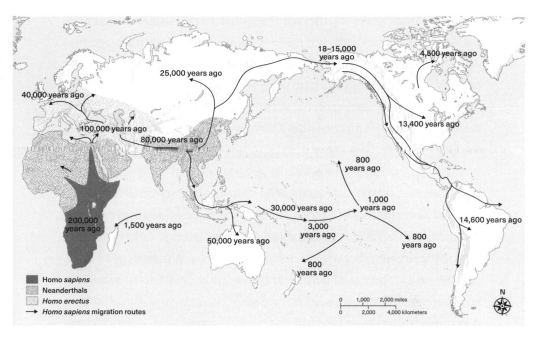

The spread of modern humans worldwide, from 100,000 years ago, moving from Africa to Europe, and across Asia to Indonesia and Australia, and northwards across the Bering Strait to the Americas.

BEING AN ANTHROPOLOGIST

A key figure in this story was the legendary Louis Leakey (1903–1972), himself born in Kenya and fluent in Kikuyu, the local language. He studied at the University of Cambridge and kept returning to Africa to collect fossils, including at famous sites such as Olduvai, where he and colleagues found numerous examples of the early human-like *Australopithecus*, with its small brain and upright gait. Leakey lived a colourful life, rarely holding a traditional academic post, and so he found ways to support himself and his family by writing books, delivering lectures and other means – for a time, he was employed as a policeman in Kenya. Although the road to independence from British rule was messy, the Leakeys were Kenyan citizens, so they maintained a life in the country after independence.

The Leakey family were and are all palaeoanthropologists, each with their independent discoveries. Mary Leakey (1913–1996), wife of Louis, made great contributions in discoveries of *Proconsul* and research on *Australopithecus*, and a remarkable site with footprints of these early humans. Their son Richard Leakey (1944–2022) was equally famous for his work on early humans, as well as for holding important roles, as Director of the National Museum of Kenya, and later as Chairman of the Kenya Wildlife Service. There he fought successfully to stop elephant poaching for their ivory, but was injured in an aircraft accident in 1993 and lost both legs below the knee. Sabotage was suspected, but never proven. He would visit the University of Bristol in the early 1990s to deliver and retrieve his daughter Louise, herself a noted palaeoanthropologist. The year after the accident, we were sitting in the White Bear, a fine old pub on campus, and before quaffing his beer, Richard removed his tin legs and stood them to one side, saying how they chafed. A redoubtable figure like his parents, Richard was not fazed by losing his legs or by the occasional dangers of Kenyan politics.

Over the years, both father and son had many run-ins with other palaeoanthropologists, and who can say who was in the right? Because of their Kenyan citizenship and senior positions in the Kenyan government, they could control access by others to the museum and to fossil sites. It seems that palaeoanthropology has always been a contentious field, often, it's said, because there are as many scientists as there are fossils. As the field of human evolution is about our own ancestry, the stakes are high when it comes to pronouncing on origins, intelligence and the human race. Two hopeful points are first, that many palaeoanthropologists manage not to squabble, and second, whatever the personal animosities might be, the science depends on the evidence, so the truth will emerge through the usual, rigorous procedures for formal scientific publication.

THE FROZEN NORTH

Much of human evolution took place against a background of reducing temperatures. This led to the Ice Ages, a feature of the

Pleistocene epoch, which lasted from 2.58 million years to 11,700 years ago. This was the time of mammoths, woolly rhinos, cave lions and Neanderthals (see Plate XXVI). What caused it and what is the legacy today?

At maximum extent, there were huge ice sheets over the northern continents, extending across most of Canada and down past Chicago in the United States, over northern Europe as far south as London and Berlin, and over much of Russia. All high areas to the south, such as the Alps and Himalayas, as well as mountains in South America, Africa, Australia and New Zealand, had larger ice caps than they do today. The great volumes of ice led to lowered sea levels – for example, the UK was linked to continental Europe, so early humans as well as the ice-loving mammals could walk back and forwards over what is now a 100-km (60-mile) strip of sea.

Great ice sheets extended south from the North Pole during the Pleistocene, covering Canada and the northern United States, as well as northern Europe and Asia.

Over the past million years, climates switched between eleven major cold phases, called glacials, and warmer interglacials. While mammoths and cave bears were widespread during the glacials, as the ice sheet shrank north at the beginning of an interglacial, climates sometimes became much warmer than today, and hippopotamuses and elephants moved north as far as London. Famously, about 120,000 years ago, herds of hippopotami colonized Trafalgar Square, as was discovered during roadworks in 1957. Just as there had been a drop of 5°C (9°F) in air and ocean temperatures 34 million years ago, there was another similar cooling event 14 million years ago, just at the time of the Fort Ternan mammals, and then global temperatures nosedived again at the end of the Pliocene and beginning of the Pleistocene. The exact cause is not certain, but the cyclicity of glacials and interglacials has long been understood.

The seemingly complex interchanges in temperature can be divided into three astronomical parameters, named Milankovitch cycles, after the Serbian mathematician Milutin Milankovitch (1879–1958) who first identified them. The cycles have spacings of 100,000 years, 41,000 years and 23,000 years, generated respectively by the eccentricity of the Earth's orbit around the Sun, the tilt of the north–south axis of the Earth and the slow wobble of the Earth on its axis, termed precession. Thanks to these three aspects of the Earth's orbit and axial wobble, summer and winter vary in length or the planet varies in distance from the Sun; when it is closer, things warm up, and when it is further away, glacials arise because of the extended darkness and cold.

MAMMOTH BLOODBATH (OR NOT)

The last Pleistocene ice age ended 11,7000 years ago, and this coincided with accelerated extinctions of many of the cold-adapted animals of northern Europe, Siberia and Canada, such as mammoths and mastodons, woolly rhinos, cave bears and Neanderthal humans with their stocky builds, heavy brows and habit of wearing animal skins to fend off the cold. Were the end-Pleistocene extinctions all a result of rapid climate change, or did the rise of humans play a part?

There has been a long and sometimes heated debate about what caused the extinction of the Pleistocene mammals: climate change vs. overkill. As evidence for overkill, some palaeontologists point to mammoth kill sites, where early humans killed and butchered the Ice Age elephants. Archaeologist Todd Surovell at the University of Wyoming has studied the La Prele Mammoth Site in Wyoming, which shows the skeleton of a subadult Columbian mammoth (Mammuthus columbi) in association with an early human camp site, dated as 12,850 years old. Surovell estimates that one mammoth would produce many tonnes of meat, equivalent to 2 million kilocalories, enough to feed thirty people for more than a month. Killing a mammoth would be very beneficial for such early peoples, but how did they do it, and how did they process the vast carcass?

Surovell reports evidence from across Europe and North America that early hunters constructed highly effective spears from wood, tipped with sharpened flint points. Many imaginative stories, images and film sequences suggest that a group of perhaps ten hunters would fan out around a likely prey animal. The elephant would be alert to danger, and although it could not run fast, it could readily kill the hunters if they got too close. It's assumed they would harry the poor beast by throwing spears, hoping to hit an artery from which gallons of blood would spurt or cut vital structures such as the hamstrings on the backs of its legs. Either way the animal might bleed to death or be unable to move, and death could take several hours.

Then, Surovell argues, the hunting band would move their camp to the carcass, rather than attempt to drag away the 5-tonne behemoth. The La Prele site shows human artefacts such as stone spear points, choppers, scrapers and even needles close to the skeleton. Perhaps the whole band squatted there, participating in the dismemberment and cooking of the meat. Some noble and strong hunters would have to chop through the shaggy pelt of the animal and its 3-cm (1-in.)-thick skin. Removing the internal organs such as the heart, liver and kidneys would require an even greater sacrifice by one of the hunters, who would hack into the belly, then climb entirely inside to cut out these favoured pieces of meat. The haunches and chops of meat would be sliced by hacking with stone choppers. Whether the hunters ate the meat or internal organs raw, as many modern hunters do, or cooked it, is not clear. Some at least was cooked, as attested by the discovery here, and at other mammoth kill sites, of broken and charred mammoth bones, some of them showing cut marks where early peoples had slashed the flesh from the bone.

Did hunting scenes such as these mark the end of the Pleistocene mammals? The overkill hypothesis has been championed by some influential researchers, but critics point out that humans were relatively rare in all parts of the world, and that different mammals went extinct at different times on different continents, sometimes coinciding with the arrival of humans, and sometimes not. Pleistocene experts had long recognized how warming climates across northern continents changed the typical vegetation from the small, cold-adapted plants that formed the basic diet of mammoths, woolly rhinos and other ice age mammals of the frozen tundra to the modern plant zones of forests, grasslands and tundra that support deer, moose and reindeer.

In a critical test between the climate change and overkill hypotheses in 2021, Mathew Stewart and colleagues of the Max Planck Institute for Chemical Ecology at Jena in Germany, showed that the most plausible case could be made for rapid climate change. They gathered data from 521 well-dated Pleistocene large mammal (megafauna) sites in North America and compared these with the detailed records of climate change and timings of the arrival of humans in different areas. The question was whether they could detect greater impacts on the population sizes of mammoths, mastodons, woolly rhinos and the others either from the arrival of humans or from sharp changes in temperature. Climate won each time. Doubtless hunting played a part, but these people did not have shotguns, nor we assume did they hunt indiscriminately, so rapid climate change and competition with more warm-adapted animals probably played a larger part in their end-Pleistocene extinction.

This extinction event marks the point at which humans began to evolve characteristics of civilization, such as complex languages, tools, art and religions. From the time of the La Prele mammoth kill site in Wyoming, humans have interacted with the natural world, and often in ways that have damaged it – in the final chapter, we explore extinction in this time of human civilization and compare it with that of the geological past, and assess where we are today and where we may be going in the future.

CHAPTER 15

The Industrial Age

SLAUGHTER OF THE DODOS

Imagine you were the first person to see a dodo – what did it look like? We may be familiar with Lewis Carroll's stuffy, besuited, professor-like bird, that waddled about, calling the other animals to order in *Alice's Adventures in Wonderland* (1865). But this dodo was autobiographical, modelled on Charles Dodgson himself, who led a double life as a Lecturer in Mathematics at the University of Oxford and as Lewis Carroll, the famous children's author. Carroll's description, beautifully rendered in illustrator John Tenniel's famous engraving, is not, however, what the dodo actually looked like.

Although Tenniel's drawing was based on images from zoological treatises, these were essentially wrong. As was the way then, the professors writing the descriptions of the extinct creature had never actually seen a living dodo and could not travel to the island of Mauritius in the Indian Ocean where it uniquely occurred. There are, however, various accounts written by sailors and travellers who had seen it in the seventeenth century. Some of the few drawings by eyewitnesses show quite a scraggy bird, not the globular form we are familiar with. Andrew Kitchener, a zoologist at the National Museums of Scotland, estimated its true weight was

The dodo in *Alice's Adventures in Wonderland*, speaking in learned fashion to Alice, and organizing the activities of the other animals. But how realistic is the reconstruction?

10.6–17.5 kg (23–39 lb), with males heavier than females (the popular image of the dodo makes him twice as heavy).

The dodo was a flightless pigeon, much larger than the regular town pigeon of today, but a pigeon nonetheless. Adult specimens were up to 1 m (3 ft) tall and bore an attractive plumage that was generally greyish or brownish, with lighter primary feathers, vestigial wings, and a tuft of curly tail feathers. The head was grey and naked, a feature of pigeons, and the beak was green, black and yellow. The legs were yellowish in colour. The birds used their massive, curved beaks to crack nuts and ate a variety of nutritious seeds and fruits. They may have eaten crabs and shellfish, but we don't know for sure. In fact, all that survives of the dodo are numerous bones, but no complete skeleton, and some scraps of skin and feathers in a museum in Oxford in the UK.

The dodo was first reported in 1598 and it was extinct by 1662 – a fact that illustrates the power of humans to wreak havoc, although it is debated whether these poor birds were literally all clubbed to death or died for a variety of reasons. The sailors found the dodos easy prey because they simply stood around looking at the visitors with curiosity and did not learn to flee. Although numerous sailors' accounts reported that the flesh was not particularly tasty – in 1634, for example, the English traveller Sir Thomas Herbert wrote, 'It is reputed more for wonder than for food, greasie stomackes may seeke after them, but to the delicate they are offensive and of no nourishment' – the sailors still killed many of them. Others were brought on board ship and fed as pets or brought back to Europe as curiosities. Their extinction may relate partly to the slaughter, but was perhaps more due to the depredations of animals introduced by the sailors, including pigs, cats, dogs, rats and even macaque monkeys; the rats may well have eaten dodo eggs, for example. Whatever the case, the dodos were all killed by human intervention within sixty-four years of the first report.

Were dodos at risk anyway? Yes and no. We can perhaps identify two lessons to learn from the death of the dodo, one moral and one biological. First, the moral. The dodo was a unique and extraordinary bird and its disappearance is a case of pointless extinction. There was no natural disaster, climate change or competition with other species that forced the dodo to give way. In all the natural cases of extinction of single species, or large numbers of species, there is a natural cause, and something eventually takes the place of the species that have disappeared. Most thinking people would agree that humans should not slaughter animals to extinction.

The second, biological, lesson is that species that live in restricted areas like islands and at small population sizes are at risk. This is a necessary conclusion from the long-established Species–Area Relationship in ecology, which states that the number of species in any area depends on the size of that area. Large island, lots of species; small island, not many species. This commonplace observation had been made by travellers and biologists since the early days of biogeography, and remarkably, was presented in mathematical form by the Swedish physicist Svante Arrhenius (1859–1927) in 1920.

Arrhenius made many important discoveries in physical chemistry, winning the Nobel Prize in 1903, and then, as many physicists and chemists do, turned his attention later in life to biology. He was successful in comparison to some others who thought they could mould biology, and especially evolution, into strictly mathematical form. It is not a bad endeavour, because everyone – not just scientists – likes to think there is some predictability in nature. But, whereas we can model precisely the rate of descent of an object under gravity, or the power needed to lift a large aeroplane from the ground, biology frustratingly doesn't follow many rules.

The Species–Area Relationship, however, has been identified as a reasonably constant law. So much so that the brilliant biologists Robert MacArthur and Ed Wilson extended Arrhenius's vision to a whole corpus of numerical thinking in their Theory of Island Biogeography. They showed that the relationship is true, although it is hugely modified by latitude – a physics-like equation can't be formulated that says, for example, 'take an island of a square kilometre or a hundred square kilometres, and these are the numbers of plant and animal species you'll find'. In the tropics, the island might sport a hundred species, but in Arctic realms, only two or three. Prediction is difficult.

MacArthur and Wilson also noted that the number of species on your island depends on how far it is from the nearest mainland. Mauritius lies about 800 km (500 miles) from Madagascar, also an island, and 1,700 km (1,055 miles) from the African mainland. This means there have been fewer opportunities for species to reach Mauritius than an island that perhaps lay only 10–20 km (6–12 miles) offshore. Plants can spread across oceans by blowing seeds and spores, or even be carried on the feet of waterbirds. Animals get to islands in different ways: by wind if they fly (birds, bats, insects) or are small (spiders, insects); or by rafting – floating by accident in clumps of timber and soil washed out to sea by storms.

Weaving these observations together, ecologists have shown time and again that biodiversity anywhere depends on temperature and food supply, and that food supply on land depends on rainfall. So, in fact, modelling a combination of average temperature and rainfall can give a reasonable prediction of species diversity at any spot on land (see Chapter 5). In the oceans, the modelling combination is temperature and productivity (the amount of food, generally measured as the volume of plankton per cubic metre of water). Conservation biologists identify hotspots as areas on land and in the oceans where biodiversity is unusually high. These biodiversity hotspots, such as tropical rainforests and coral reefs, all lie around the Equator.

The other main species risk factor is dietary specialization. Pandas eat solely young bamboo shoots so struggle to find enough nutrition even when conditions are good, and this is why they are such an emblem of extinction risk. Dodos, so far as we can tell, ate a broad diet of nutritious plant material (fruits and seeds are always better than plant stems), so could presumably survive in varying conditions.

Biology, then, tells us that specialization and small population sizes are serious risk factors for extinction. Better not be a species that occurs on only one island or mountain top, and better not become too specialized in your diet or other life requirements. The conservation Red List of species at risk of extinction contains many such species. Better to be a cockroach or a pigeon … or is that always the case?

THE TRAGEDY OF THE PASSENGER PIGEON

Pigeons seem to be everywhere, especially if you live in a town. They are legally classed as vermin in many parts of the world, meaning you can freely shoot them on farmland. In London, New York and Cape Town, they hop around on the streets foraging for crusts and other debris, knocked about by the traffic, but seemingly indestructible. They are widespread, there are hundreds of millions of them and they seem hugely adaptable to all weathers and food supplies. The most abundant native North American pigeon in the early 1800s, however, was the passenger pigeon, which lived in the deciduous woodlands around the Great Lakes region, where it fed on forest fruits and seeds like acorns, nested communally and sometimes migrated in huge flocks of millions of birds in search of food. It was estimated that there were three to five billion passenger pigeons in North America. The great ornithologist James John Audubon wrote in 1813, as he watched a huge flock pass by, 'The air was literally filled with Pigeons; the light of noon-day was obscured as by an eclipse; the dung fell in spots, not unlike melting flakes of snow, and the continued buzz of wings had a tendency to lull my senses to repose.'

Early farmers in North America reported that these vast flocks blackened the sky and filled them with fear because, like a flock of locusts, the huge numbers of birds could devour entire fields of grain crops. Although Audubon appreciated them as a wonder of nature, the new farmers did not. The birds were hunted by native Americans and European settlers, but with their guns and concentrated hunting parties, farmers could kill large numbers and this was considered essential to protect their crops. Some rare voices urged caution and reminded their fellow citizens that the pigeons might not always be there. Despite conservation laws in the 1870s and 1880s following massive reductions in flock sizes, the slaughter continued. Farmers and their guns would not be separated. Bags of 30,000 or 50,000 birds were proudly touted and rewarded with prizes; the country folk were not killing the pigeons for food, but for fun.

The tragedy could have been reversed in the 1870s, when there were still roosts of several thousand birds, but the last wild bird was shot in 1901. By then it was too late, even though there were still dozens of specimens in zoos; the final passenger pigeon, called Martha, breathed her last in Cincinnati Zoo in 1914. Conservation biologists talk about a Minimum Viable Population (MVP), the smallest number of individuals of a species required to ensure its survival. The MVP is more than two, and in fact considerably more than two. The Noah's Ark model of saving a male and female of each bird, mammal and reptile is inadequate because those two do not sample the full breadth of the gene pool of the species.

Usually, the correct MVP for birds and mammals is 500–1,000, and even those figures involve the risk of inbreeding, where closely related animals cross-breed, such as child and parent, or close cousins. An MVP of more than 5,000 is much better, ideally including samples of two or three geographically separated groups – for example, passenger pigeons from Michigan, New York and Pennsylvania. That's why efforts by zoos to save species from starting populations of five or six are tricky and require much hard work. In the wild, such low numbers would indicate that the species is doomed.

Similar slaughter was wrought on the North American bison, once so common that the first Europeans to see their herds in the

Bison skulls. This photograph forces a double-take; are there really 10,000 bison skulls in this pile? European settlers in North America in the 1800s were enthusiastic killers, and here they were not even eating the bison flesh.

early 1800s were amazed by such richness and productivity. If bison by the thousands could graze the North American grasslands, these must be excellent lands for farming. Fortunately, the shooting stopped before all bison had been killed and they are now proudly conserved by modern citizens of North America.

Are Europeans and non-native Americans always the bad guys? Not necessarily.

THE MOAS AND THE MAORI

Some of the first recorded human-caused extinctions occurred in the 1400s in New Zealand, which used to be home to some amazing flightless birds, the moas. There were nine species, the largest of which was 3.6 m (12 ft) tall and the smallest about the size of a turkey. These relatives of modern ostriches and emus had become flightless millions of years ago, perhaps because New Zealand, having been isolated from other parts of the world for a long time, had no native mammals, therefore various birds took over mammal-like roles. There are still flightless parrots acting as predators (maybe standing in for wild cats) and also flightless rails, flightless ducks and, of course, the kiwis, chicken-sized nocturnal hunters that presumably take the place filled by hedgehogs and rats in other lands.

The moas reached New Zealand 60 million years ago, when the islands were still connected to Australia, and then diversified, feeding on leaves and twigs, and standing in ecologically for antelope and llamas. Polynesian peoples arrived in New Zealand some time before 1300 and settled, becoming the Maori of today. They hunted the moas for food, and, to some extent, their forest clearances may have contributed to the death of the species. As far as is known, all nine species of moa were extinct by 1445, only 150 years after the Maori settled.

Unlike the cases of the dodo and passenger pigeon, however, there is no evidence the Maori practised overhunting (in other words, killing animals because they are easy to kill). They were presumably killed specifically for food, but the new human predators were too effective and too quick, and the moas could not evolve the necessary behaviour and habits fast enough to enable them to survive. Moas had been preyed on by the remarkable Haast's eagle, twice the size of most modern eagles, but it, too, went extinct around 1400. The conclusion is not that the Maori were wilfully destructive, but that humans, just by their existence, cause extinction.

So, we've learned that species are at risk by being overly specialized or living in low numbers, as was the case for the dodo, and possibly also for the moas. Island life has always been a danger, especially in cases such as New Zealand, where a full complement of species is not present and predation is non-existent or low. Also, it's clear that humans seem to have a negative effect, especially on islands, where the native species may not be used to human behaviour: even if we try to live in harmony with nature, we still kill for food and can destroy habitats without realizing what we are doing.

HOLOCENE AND ANTHROPOCENE

In our adventure through deep time, we have spanned over 600 million years from the Ediacaran animals of the late Precambrian and the Cambrian explosion through to the extinction event at the end of the Ice Ages, 11,700 years ago, which marks the end of the Pleistocene Epoch and beginning of the Holocene Epoch. Climates continued to cool from the Pleistocene, but the ice sheets had withdrawn and did not return to their former wide extent. The mammoths, cave bears and giant Irish deer had gone, although Pleistocene animals such as European bison, wolves, bears and lynx survived. These were eventually largely wiped out in Europe by long-term pressure of human populations, but bison, caribou, bears and other large mammals survive in natural wildernesses in North America.

A new geological epoch has been proposed: the Anthropocene. This is based on a term derived from the Greek *anthropos*, meaning 'human', and was used sporadically last century, but since 2000 has been adopted and promoted by the atmospheric chemist Paul Crutzen. The term is intended to reflect human influence on the Earth, which Crutzen and others felt, particularly after industrialization in the 1700s, has been so immense that it is on a par with all the previous natural crises and climate changes we have looked at. The problem with this proposal is that there is no agreement about when the Anthropocene started – was it 12,000 years ago when humans began to settle and farm; or the fourteenth century, when Europeans began large-scale global exploration and colonization; or from 1700, when coal-based industries began to develop in Europe; or 1945, when the first atomic bomb was detonated; or 1987, when total global population reached 5 billion?

Most, including Crutzen, would refer to the point at which human activity began to affect the atmosphere in a serious way (see Plate XXVII). This could be taken as the beginning of the industrial revolution, before 1800, or 1960, when the human influences began to outweigh the natural influences on carbon dioxide in the atmosphere and global temperature. There has been a strong campaign to have the Anthropocene recognized as an official geological epoch, not equivalent to the Holocene but reflecting something new about the twentieth century and the depth of human influence on the functioning of Earth's climates. Various commissions and agencies have promoted the case, but the relevant international geological commissions have been reluctant so far to move. We'll see.

HOW MUCH DO ALL HUMANS WEIGH?

In a way, the Anthropocene squabble is a sideshow, because the influence of humans on the Earth and nature is widely accepted. Ever since people grouped together in cooperative bands to hunt large mammals such as mammoths and mastodons, we have been having an effect. Although the story of the Maori and the moas shows that humans can cause widespread extinctions while living in some way in harmony with nature, the escalation in the slaughter of species and destruction of habitats from the nineteenth century onwards has been incredible, and this is only made much worse by the pressure of human population.

The statistics of human population increase are clear: 1 billion in 1800; 2 billion in 1927; 4 billion in 1974; 6 billion in 1999; 8 billion in 2022. This is an exponential increase, with the population doubling in 127 years from 1800 to 1927, and then a continuous shortening of the time to add another 2 billion people: 47 years, 25 years and 23 years. We can only hope the rate is slowing down and might peak this century, and then even start to go down. This may not always be encouraged by those in power, however, because the success of nations depends on increasing population sizes and gross domestic product, which means more forest clearance, more production and more greenhouse gases. Viewed this way, any green living measures that are adopted, like using a bicycle, eating only plant food or using recycled coffee cups, can have only a trivial effect while natural habitats are being converted to farmland at a colossal rate. However, this is not a reason not to do our bit, and I continue to walk, eat local food and recycle my waste in a vaguely optimistic way.

Can we measure the impact of humans on life? One way was suggested by Yinon Bar-On of the Weizmann Institute of Science in Israel in 2018. He and colleagues estimated the total mass of all living things on Earth and came up with the boggling figure of 550 gigatonnes of carbon. Humans make up a measly 0.06 gigatonnes, and their domesticated mammals (pigs, sheep, cattle, camels) somewhat more, at 0.1 gigatonnes. But the sum total of all wild mammals is now much less, only 0.007 gigatonnes. It's easier to comprehend this in a pie chart of all mammals: all wild mammals together are 4% of the pie, humans are 36%, and our domesticated mammals make up 60%. The same authors estimate that farmed chickens make up three times the total biomass of all ten thousand species of wild birds.

ANTHROPOGENIC CLIMATE CHANGE

Most debated and disputed, of course, have been the greenhouse gases. We're talking mainly about carbon dioxide, which has been such an important atmospheric gas throughout the history of the Earth. Recall that its increase was the primary driver during hyperthermal events, whether large like the end-Permian mass extinction or much smaller like the Palaeocene–Eocene Thermal Maximum (PETM). Levels of carbon dioxide through the Mesozoic and Cenozoic have determined global temperatures throughout, and a decrease in carbon dioxide has been shown to be the main cause, for example, of the down-stepping global temperatures on Earth from 34 million years ago to the present day.

Everyone has witnessed the unedifying climate change sceptic debates played out in the media. These debates have been promoted by business and fundamentalist organizations in the United States and elsewhere. For example, in the US the same lobbying groups and methods that were deployed in the 1960s to try to persuade people that smoking was a great thing have been used. As the brilliant investigative historians of science Naomi Oreskes and Erik Conway showed in their 2010 book, *Merchants of Doubt: How a Handful of Scientists Obscured the Truth on Issues from Tobacco Smoke to Global Warming*, the climate change deniers in the United States dug up the same unethical and antiquated scientists who had earlier said cigarettes were an important part of a healthy lifestyle. I need not repeat all the clearly articulated arguments for climate change given in their book and elsewhere: the current huge human population coupled with energy-intensive lifestyles is not only killing plants and animals at unprecedented rates but also generating sufficient carbon dioxide to change world climates.

What is the evidence for a warming climate? The temperature-through-time graph shows that average global temperature has gone up by more than 1°C (2°F) since 1900. The graph begins in 1850 when instrumental temperature readings began to be collected by physicists. Although the record can be extended back by thousands or millions of years by using ice cores, ancient trees, oxygen isotopes and the like, instrumental measurements are strongly reliable figures. Temperatures fluctuate year by year of course, and the pattern shows swings up and down of no more than 0.5°C (1°F) from 1850 to 1950. About 1920, human impact began to push temperatures higher than expected at some points, but this effect becomes clear after 1960. After that date, global temperatures take off, rising out of the ±0.5°C (1°F) envelope, passing a 1.0°C (2°F) increase in 2010 and heading towards 1.5°C (2.7°F) in 2030. Without the impact of human activity, the estimated world temperature increase would still be around zero.

We also see the changes in temperature regionally. Although there has been no change in small patches in the North Atlantic and in a strip around Antarctica, and there are even slight temperature falls in the Weddell Sea at the southernmost end of the South Atlantic, elsewhere temperatures are warming, especially around the North Pole, by over 4°C (7.2°F) in some places, hence the precarious position of polar bears and other animals finely tuned to living on continuous ice sheets. Note also the large temperature rises of up to 2°C (3.6°F) over most of Europe, Africa, Asia and parts of the Americas. The oceans mitigate things a little by mixing cooler waters from the Poles and so experience slightly smaller temperature rises.

THE IMPACT OF RISING TEMPERATURE

Living in northern temperate Europe as I do, it is easy to joke about the temperature rise. Yes, we experience drier summers and more serious hurricane-like storms, but the generally warmer temperatures mean we don't get heavy winters any more, and vineyards are spreading north. Who wouldn't exchange a damp and cold British climate for the balmy sunshine and gorgeous wines of the south of France? Similarly, in Chicago and New York, winters are shorter and summers longer, although hurricanes from the Caribbean are more frequent and nastier.

But spare a thought for the often voiceless peoples who live around the tropical zones. Each year, the Sahara Desert increases in size by 7,600 sq. km (3,000 sq. miles) and is, in 2022, 10% larger than it was in 1920. These are raw figures, but that is equivalent to the death of several thousand people. Whole villages migrate each year out of areas they used to farm. People in Africa, India, South China and Australia are getting used to summer temperatures over 40°C (104°F), the death temperature for most plants and animals, as we saw in Chapter 5. Air conditioning, which helps to keep populations in wealthy countries cool in the heat, is not the answer: it eats power and generates carbon dioxide, which in turn raises temperatures, contributing to driving people from their homes around the tropics.

The expansion of the Sahara Desert is just one of many such signs of rapid temperature change that also hurts wildlife, killing plant and animal species that can't escape or find another place to live. Some may argue that a temperature rise of a degree or two won't make much difference, but it does. Many of the mass extinctions of the past were driven by temperature rises of as little as 5°C (9°F), enough to shift tropical air and sea surface temperatures from comfortable into the death zone.

THE LESSONS LEARNT

Have we learnt anything else from our mad rush through geological time, stopping off to look intently at the various trials and crises life has passed through? As we have seen in discussions of the current crisis, the past really does inform the present. We can compare the present day, with its well-recorded historical changes in species numbers and global temperatures, with such data from deep time. The way the Earth, atmosphere and oceans work today is the way they have always worked. Therefore, the approach in this book, of moving back and forwards in time, looking at some extraordinary life of the past, such as the Ediacaran animals or the dinosaurs, and then comparing them with modern worms or crocodilians, is entirely appropriate. Exploring all the rich time-series data on changing oceanic and atmospheric chemistry and temperatures is another excellent way to assess what effects these many different scenarios have had on Earth function and biodiversity.

Another lesson from the past is that hyperthermal events represent a predictable model for large-scale extinction. Intense study of the 56-million-year-old PETM has confirmed it was a hyperthermal event linked to eruptions of volcanoes in the North Atlantic and the release of deep-ocean methane hydrates. The temperature rise was up to 6°C (11°F), and extinctions were serious in some habitats, but not in others. At the other end of the scale, the end-Permian mass extinction 252 million years ago was linked to temperature rises of 10–15°C (18–27°F) and the impact on biodiversity could not have been more severe: 90–95% of species disappeared. The universal hyperthermal model of eruption, carbon dioxide release, global warming, acid rain, ocean acidification and ocean anoxia seems to work for a dozen or more past events, and it also seems to be working for the current hyperthermal crisis. It's not volcanic eruptions, of course, that are now pumping carbon dioxide into the atmosphere, but cars, factories, planes, cows and all kinds of other producers that serve humanity. All these factors are orchestrated by human needs and desires, and they produce huge amounts of carbon. Temperatures continue to rise, and all the consequences follow. At least with the past extinction events, the production of carbon dioxide eventually came to an end as volcanic eruptions shut down, allowing the atmosphere and oceans to return to normal after some hundreds or thousands of years.

As we have seen, life has always recovered from devastating past crises, so it might be possible for an optimist to argue, 'well, life always found a way and recovered' or 'the Earth can bounce back from carbon dioxide shocks'. While both statements are true, we can also see that there is a problem with timescales: the recovery of climate equilibrium might take a thousand years, and the evolutionary bursts of new life after mass extinctions can take a million years or more. Good for the Earth and life, but not for humans with our short lives and short attention spans.

This is emphasized by what we have seen in this chapter about the extraordinary effect humans have had on driving extinction, which is a new phenomenon and does not relate to any of the deeper time mass extinctions or extinction events. The human impulse to kill, usually causing extinction inadvertently, is problematic. Before humans, extinction was a part of nature, something that fitted with the multiple short-, medium- and long-term processes that perturb the Earth and life. Unlike these natural processes, which include shockingly massive volcanic eruptions and asteroid impacts, however, humans are aware of their actions and can choose what to do and what not to do. So why should we avoid causing extinction? As I noted at the beginning of this book, E. O. Wilson called the assemblage of ethical arguments 'biophilia', meaning a love of life. Chief among these is that nature is beautiful and remarkable, and every species has value; we regret if a species like the dodo is no longer there to be seen. Perhaps we don't regret the extinction of a species of cockroach, seaweed or mosquito, but we should.

Finally, in drawing together the threads of our exploration of extinction events, the evidence shows that the end-Precambrian events enabled the origin of modern animal groups in the Cambrian explosion; the end-Cambrian mass extinction triggered the Great Ordovician Biodiversification Event, when so many marine groups expanded and life crept onto land; the end-Permian mass extinction triggered the Triassic Revolution, with all those delicious, meaty sea creatures and insulated, warm-blooded proto-dinosaurs and proto-mammals; and the end-Cretaceous mass extinction gave angiosperms, mammals and birds their chances to flourish. Other events, such as the PETM and the end-Pleistocene extinctions, seem to have coincided more with origins and opportunities than with the devastating loss of huge numbers of species. Through this review, a novel finding has been revealed: these past extinction events, viewed from the present-day, all made remarkably creative contributions to the overall history of life.

Eon   **Era**

PHANEROZOIC

CENOZOIC
··66 ···································

MESOZOIC
··252 ·······························

PALAEOZOIC

··541 ·······························

PROTEROZOIC
Bacteria, algae, jellyfish

2,500

ARCHEAN
Earth's crust has cooled
enough to allow the
formation of continents
and life starts to form

4,000

HADEAN
Formation of the Earth

4,540 million years ago

Era	Period	
		Present day

CENOZOIC
- **Quaternary** Rise of humans — 1.8
- **Tertiary** Rise of mammals — 66

MESOZOIC
- **Cretaceous** Modern seed-bearing plants, dinosaurs — 145 — **End-Cretaceous extinction**
- **Jurassic** First birds — 201
- **Triassic** Cycads, first dinosaurs — 252 — **End-Triassic extinction**

PALAEOZOIC
- **Permian** First reptiles — 299 — **End-Permian extinction**
- **Carboniferous** First insects — 359
- **Devonian** First seed-bearing plants, cartilaginous fishes — 419 — **End-Devonian extinction**
- **Silurian** First land animals — 444
- **Ordovician** Early bony fish — 485 — **End-Ordovician extinction**
- **Cambrian** Invertebrate animals, brachiopods, trilobites — 541 million years ago

Bibliography

PREFACE

— Rosenzweig, M. L. and McCord, R. D. 1991. 'Incumbent replacement: evidence for long-term evolutionary progress'. *Paleobiology* 17, 23–27.
— Shapiro, B. 2015. *How to Clone a Mammoth*. Princeton University Press, Princeton.
— Slater, G. J. 2013. 'Phylogenetic evidence for a shift in the mode of mammalian body size evolution at the Cretaceous-Palaeogene boundary'. *Methods in Ecology and Evolution* 4, 734–44.
— Wilson, E. O. 1984. *Biophilia*. Harvard University Press, Cambridge, Mass.
— Wilson, E. O. 1992. *The Diversity of Life*. Harvard University Press, Cambridge, Mass.
— Wrigley, C. A. 2021. 'Ice and ivory: the cryopolitics of mammoth de-extinction'. *Journal of Political Ecology* 28, 782–803.

---

1. THE FIRST ANIMALS AND MASS EXTINCTIONS

— Hoffman, P. F. and Schrag, D. P. 2000. 'Snowball Earth'. *Scientific American* 282(1), 68–75.
— Lyons, T., Reinhard, C. Y. and Planavsky, N. J. 2014. 'The rise of oxygen in Earth's early ocean and atmosphere'. *Nature* 506, 307–15.
— Retallack, G. J. 1994. 'Were the Ediacaran fossils lichens?'. *Paleobiology* 20, 523–44.
— Schopf, J. W. 2021. 'Precambrian paleobiology: precedents, progress, and prospects'. *Frontiers in Ecology and Evolution* 9, 707072.
— Seilacher, A. 1989. 'Vendozoa: organismic construction in the Proterozoic biosphere'. *Lethaia* 22, 229–39.

— Seward, A. C. 1931. *Plant Life Through the Ages*. Cambridge University Press, Cambridge.
— Sprigg, R. C. 1947. 'Early Cambrian "jellyfishes" of Ediacara, South Australia and Mount John, Kimberly District, Western Australia'. *Transactions of the Royal Society of South Australia* 73, 72–99.
— Tyler, S. A. and Barghoorn, E. S. 1954. 'Occurrence of structurally preserved plants in Precambrian rocks of the Canadian shield'. *Science* 119, 606–8.
— Walcott, C. D. 1899. 'Precambrian fossiliferous formations'. *Bulletin of the Geological Society of America* 10, 199–244.

2. THE CAMBRIAN EXPLOSION AND EXTINCTIONS

— Budd, G. 2013. 'At the origin of animals: the revolutionary Cambrian fossil record'. *Current Genomics* 14, 344–54.
— Conway Morris, S. *The Crucible of Creation: The Burgess Shale and the Rise of Animals*. Oxford University Press, Oxford.
— Darroch, S. A. F., Smith, E. F., Laflamme, M. and Erwin, D. H. 2018. 'Ediacaran extinction and Cambrian explosion'. *Trends in Ecology and Evolution* 33, 653–63.
— Erwin, D. H., Laflamme, M., Tweedt, S. M., Sperling, E. A., Pisani, D. and Peterson, K. J. 2011. 'The Cambrian conundrum: early divergence and later ecological success in the early history of animals'. *Science* 334, 1091–97.
— Erwin, D. H. and Valentine, J. W. 2013. *The Cambrian Explosion: The Construction of Animal Biodiversity*. Roberts and Company Publishers Inc., Greenwood Village, Colo.
— Gould, S. J. 1989. *Wonderful Life: The Burgess Shale and the Nature of History*. W.W. Norton, New York.
— Knoll, A. H. and Carroll, S. B. 1999. 'Early animal evolution: emerging views from comparative biology and geology'. *Science* 284, 2129–37.
— Whittington, H. B. 1985. *The Burgess Shale*. Yale University Press, New Haven.

3. ORDOVICIAN DIVERSIFICATION AND MASS EXTINCTION

— Bancroft, B. B. 1933. *Correlation Tables of the Stages Costonian–Onnian in England and Wales*. Published by the author, Blakeney, Glos.
— Bancroft, B. B. 1945. 'The brachiopod zonal indices of the stages Costonian to Onnian in Britain'. *Journal of Paleontology* 19, 181–252.
— Cocks, L. R. M. 2019. *Llandovery Brachiopods from England and Wales*. *Monographs of the Palaeontographical Society* 172, 1–262.
— Elles, G. L. 1922. 'The age of the Hirnant Beds'. *Geological Magazine* 59, 409–14.

— Harper, D. A. T. 2021. 'Late Ordovician extinctions'. In *Encyclopedia of Geology (Second Edition)*. Elsevier, London, pp. 617–27.
— Harper, D. A. T., Hammarlund, E. U. and Rasmussen, C. M. Ø. 2014. 'End Ordovician extinctions: a coincidence of causes'. *Gondwana Research* 25, 1294–1307.
— Harper, D. A. T., Zhan, R. B. and Jin, J. 2015. 'The Great Ordovician Biodiversification Event: reviewing two decades of research on diversity's big bang illustrated by mainly brachiopod data'. *Palaeoworld* 24, 75–85.
— Jones, O. T. 1923. 'The Hirnant Beds and the base of the Valentian'. *Geological Magazine* 60, 514–19.
— Lamont, A. 1946. 'Mr B. B. Bancroft' [obituary]. *Nature* 157, 42.
— Ling, M. X., Zhan, R. B., Wang, G.X., et al. 2019. 'An extremely brief end Ordovician mass extinction linked to abrupt onset of glaciation'. *Solid Earth Sciences* 4, 190–8.
— Longman, J., Mills, B. J. W., Manners, H. R., Gernon, T. M. and Palmer, M. R. 2021. 'Late Ordovician climate change and extinctions driven by elevated volcanic nutrient supply'. *Nature Geoscience* 14, 924–29.
— Rong, J. Y., Harper, D. A. T., Huang, B., Li, R. Y., Zhang, X. L. and Chen, D. 2020. 'The latest Ordovician Hirnantian brachiopod faunas: new global insights'. *Earth-Science Reviews* 208, 103280.
— Sepkoski, J. J., Jr. 1984. 'A kinetic model of Phanerozoic taxonomic diversity. III. Post-Paleozoic families and mass extinctions'. *Paleobiology* 10, 246–67.
— Servais, T. and Harper, D. A. T. 2018. 'The Great Ordovician Biodiversification Event (GOBE): definition, concept and duration'. *Lethaia* 51, 151–64.
— Tubb, J. and Burek, C. 2020. 'Gertrude Elles: the pioneering graptolite geologist in a woolly hat. Her career, achievements and personal reflections from her family and colleagues'. In *Celebrating 100 Years of Female Fellowship of the Geological Society: Discovering Forgotten Histories* (ed. C. V. Burek and B. M. Higgs), Special Publications of the Geological Society 506, pp. 157–69.

4. THE MOVE TO LAND AND THE LATE DEVONIAN CRISIS

— Abbasi, A. M. 2021. 'Evolution of vertebrate appendicular structures: insight from genetic and palaeontological data'. *Developmental Dynamics* 240, 1005–16.
— Anderson, P. S. L. and Westneat, M. W. 2007. 'Feeding mechanics and bite force modelling of the skull of *Dunkleosteus terrelli*, an ancient apex predator'. *Biology Letters* 3, 77–80.
— Carr, R. K. 2010. 'Paleoecology of *Dunkleosteus terrelli* (Placodermi, Arthrodira)'. *Kirtlandia* 57, 36–45.
— Carr, R. K. and Jackson, G. L. 2008. 'The vertebrate fauna of the Cleveland Member (Famennian) of the Ohio Shale'. *Ohio Geological Survey Guidebook* 22, 1–17.

– Clack, J. A. 2012. *Gaining Ground: The Origin and Evolution of Tetrapods* (2nd edn). Cambridge University Press, Cambridge.
– Coates, M. I. 2003. 'The evolution of paired fins'. *Theory in Biosciences* 122, 266–87.
– Friedman, M. and Sallan, L. C. 2012. 'Five hundred million years of extinction and recovery: a Phanerozoic survey of large-scale diversity patterns in fishes'. *Palaeontology* 55, 707–42.
– Gensel, P. G., Glasspool, I., Gastaldo, R. A., Libertín, M. and Kvaček, J. 2020. 'Back to the beginnings. the Silurian Devonian as a time of major innovation in plants and their communities'. In *Nature Through Time* (ed. E. Martinetto, E. Tschopp and R. A. Gastaldo). Springer, Cham, Switzerland, pp. 367–98.
– Kaiser, S. I., Aretz, M. and Becker, R. T. 2015. 'The global Hangenberg Crisis (Devonian-Carboniferous transition): review of a first-order mass extinction'. In *Devonian Climate, Sea Level and Evolutionary Events* (ed. R. T. Becker, P. Königshof and C. E. Brett), Special Publications of the Geological Society 423, pp. 387–437.
– McGhee, G. R., Jr. 2013. *When the Invasion of Land Failed: The Legacy of the Devonian Extinctions*. Columbia University Press, New York.
– Sallan, L. C. and Coates, M. I. 2010. 'End-Devonian extinction and a bottleneck in the early evolution of modern jawed vertebrates'. *Proceedings of the National Academy of Sciences, USA* 107, 10131–35.
– Servais, T., Cascales-Minana, B., Cleal, C. J., Gerrienne, P., Harper, D. A. T. and Neumann, M. 2019. 'Revisiting the great Ordovician diversification of land plants: recent data and perspectives'. *Palaeogeography, Palaeoclimatology, Palaeoecology* 534, 109280.

5. HOW GLOBAL WARMING KILLS

– Bartels, D. and Hussain, S. S. 2011. 'Resurrection plants: physiology and molecular biology'. *Ecological Studies* 215, 339–64.
– Bedford, T. 1951. 'Obituary: H. M. Vernon, D.M.'. *British Medical Journal* 1951(1), 419.
– Benton, M. J. 2018. 'Hyperthermal-driven mass extinctions: killing models during the Permian-Triassic mass extinction'. *Philosophical Transactions of the Royal Society, Series A* 376, 20170076.
– Day, M. O. and Rubidge, B. S. 2021. 'The late Capitanian mass extinction of terrestrial vertebrates in the Karoo Basin of South Africa'. *Frontiers in Earth Science* 9, 631198.
– Jagadish, S. V. K., Way, D. A. and Sharkey, T. D. 2021. 'Plant heat stress: concepts directing future research'. *Plant, Cell & Environment* 44, 1992–2005.
– Jørgensen, L. B., Ørsted, M., Malte, H., Wang, T. and Overgaard, J. 2022. 'Extreme escalation of heat failure rates in ectotherms with global warming'. *Nature* 611, 93–98.

— Pörtner, H. O., Langenbuch, M. and Michaelidis, B. 2005. 'Synergistic effects of temperature extremes, hypoxia, and increases in CO_2 on marine animals: from Earth history to global change'. *Journal of Geophysical Research* 110, C09S10.
— Sahney, S., Benton, M. J. and Falcon-Lang, H. J. 2010. 'Rainforest collapse triggered Pennsylvanian tetrapod diversification in Euramerica'. *Geology* 38, 1079–82.
— Sejian, V., Bhatta, R., Gaughan, J. B., Dunshea, F. R. and Lacetera, N. 2018. 'Review: adaptation of animals to heat stress'. *Animal* 12(s2), s431–s444.
— Teskey, R., Wertin, T., Bauweraerts, I., Ameye, M., McGuire, M. A. and Steppe, K. 2015. 'Response of tree species to heat waves and extreme heat stress'. *Plant, Cell & Environment* 38, 1699–1712.
— Vernon, H. M. 1899. 'The death temperature of certain marine organisms'. *Journal of Physiology* 25, 131–36.
— Zhang, B., Wignall, P. B., Yao, S., Hu, W. and Liu, B. 2021. 'Collapsed upwelling and intensified euxinia in response to climate warming during the Capitanian (Middle Permian) mass extinction'. *Gondwana Research* 89, 31–46.

6. THE GREATEST CRISIS OF ALL TIME

— Benton, M. J. 2015. *When Life Nearly Died*. Thames & Hudson, London.
— Benton, M. J. 2018. 'Hyperthermal-driven mass extinctions: killing models during the Permian-Triassic mass extinction'. *Philosophical Transactions of the Royal Society, Series A* 376, 20170076.
— Benton, M. J. and Newell, A. J. 2014. 'Impacts of global warming on Permo-Triassic terrestrial ecosystems'. *Gondwana Research* 25, 1308–37.
— Erwin, D. H. 1993. *The Great Paleozoic Crisis*. Columbia University Press, New York.
— Erwin, D. H. 2015. *Extinction: How Life Nearly Ended 250 Million Years Ago*. Princeton University Press, Princeton.
— Newell, A. J., Tverdokhlebov, V. P. and Benton, M. J. 1999. 'Interplay of tectonics and climate on a transverse fluvial system, Upper Permian, southern Uralian foreland basin, Russia'. *Sedimentary Geology* 127, 11–29.
— Retallack, G. J., Veevers, J. J. and Morante, R. 1996. 'Global coal gap between Permian–Triassic extinction and Middle Triassic recovery of peat-forming plants'. *Geological Society of America Bulletin* 108, 195–207.
— Ward, P. D., Montgomery, D. R. and Smith, R. H. M. 2000. 'Altered river morphology in South Africa related to the Permian-Triassic extinction'. *Science* 289, 1740–43.
— Wignall, P. B. 2001. 'Large igneous provinces and mass extinctions'. *Earth-Science Reviews* 53, 1–33.
— Wignall, P. B. and Hallam, A. 1992. 'Anoxia as a cause of the Permian–Triassic mass extinction: facies evidence from northern Italy and the western United States'. *Palaeogeography, Palaeoclimatology, Palaeoecology* 93, 21–46.

7. TRIASSIC RECOVERY

— Bambach, R. K. 1993. 'Seafood through time: changes in biomass, energetics, and productivity in the marine ecosystem'. *Paleobiology* 19, 372–97.

— Benton, M. J. 2021. 'The origin of endothermy in synapsids and archosaurs and arms races in the Triassic'. *Gondwana Research* 100, 261–89.

— Benton, M. J., Dhouailly, D., Jiang, B. Y. and McNamara, M. 2019. 'The early origin of feathers'. *Trends in Ecology & Evolution* 34, 856–69.

— Hu, S., Zhang, Q., Chen, Z.-Q., Zhou, C., Tao, L., Tap, X., Wen, W., Huang, J. and Benton, M. J. 2011. 'The Luoping biota: exceptional preservation, and new evidence on the Triassic recovery from end-Permian mass extinction'. *Proceedings of the Royal Society B*, 278, 2274–82.

— Huttenlocker, A. K. and Farmer, C. G. 2017. 'Bone microvasculature tracks red blood cell size diminution in Triassic mammal and dinosaur forerunners'. *Current Biology* 27, 48–54.

— Kubo, T. and Benton, M. J. 2007. 'Tetrapod postural shift estimated from Permian and Triassic trackways'. *Palaeontology* 52, 1029–37.

— Payne, J. L, Lehrmann, D. J., Wei, J., Orchard, M. J., Schrag, D. P. and Knoll, A. H. 2004. 'Large perturbations of the carbon cycle during recovery from the end-Permian extinction'. *Science* 305, 506–9.

— Sepkoski, J. J., Jr. 1984. 'A kinetic model of Phanerozoic taxonomic diversity. III. Post-Paleozoic families and mass extinctions'. *Paleobiology* 10, 246–67.

— Smith, R. H. M. and Botha-Brink, J. 2014. 'Anatomy of a mass extinction: sedimentological and taphonomic evidence for drought-induced die-offs at the Permo-Triassic boundary in the main Karoo Basin, South Africa'. *Palaeogeography, Palaeoclimatology, Palaeoecology*, 396, 99–118.

— Van Valen, L. M. 1982. 'A resetting of Phanerozoic community evolution'. *Nature* 307, 50–51.

— Watson, D. M. S. 1931. 'On the skeleton of a bauriamorph reptile'. *Proceedings of the Zoological Society of London* 1931, 35–98.

— Yang, Z. X., Jiang, B. Y., McNamara, M. E., et al. 2019. 'Pterosaur integumentary structures with complex feather-like branching'. *Nature Ecology & Evolution* 3, 24–30.

8. THE CARNIAN PLUVIAL EPISODE AND DIVERSIFICATION OF THE DINOSAURS

— Benton, M. J. 1983a. 'Dinosaur success in the Triassic: a noncompetitive ecological model'. *Quarterly Review of Biology* 58, 29–55.

— Benton, M. J. 1983b. 'The Triassic reptile *Hyperodapedon* from Elgin: functional morphology and relationships'. *Philosophical Transactions of the Royal Society of London, Series B* 302, 605–717.

— Benton, M. J. 1985. 'More than one event in the late Triassic mass extinction'. *Nature* 321, 857–61.
— Bernardi, M., Gianolla, P., Petti, F. M., Mietto, P. and Benton, M. J. 2018. 'Dinosaur diversification linked with the Carnian Pluvial Episode'. *Nature Communications* 9, 1499.
— Dal Corso, J., Bernardi, M., Sun, Y., et al. 2020. 'Extinction and dawn of the modern world in the Carnian (Late Triassic)'. *Science Advances* 6, eaba0099.
— Dal Corso, J., Gianolla, P., Newton, R. J., et al. 2015. 'Carbon isotope records reveal synchronicity between carbon cycle perturbation and the "Carnian Pluvial Event" in the Tethys realm (Late Triassic)'. *Global and Planetary Change* 127, 79–90.
— Ruffell, A., Simms, M. J. and Wignall, P. B. 2016. 'The Carnian Humid Episode of the late Triassic: a review'. *Geological Magazine* 153, 271–84.
— Simms, M. J. and Ruffell, A. H. 1989. 'Synchroneity of climatic change in the late Triassic'. *Geology* 17, 265–68.
— Simms, M. J. and Ruffell, A. H. 2018. 'The Carnian Pluvial Episode: from discovery, through obscurity, to acceptance'. *Journal of the Geological Society* 175, 989–92.
— Visscher, H., Van Houte, M., Brugman, W. A. and Poort, R. J. 1994. 'Rejection of a Carnian (Late Triassic) "pluvial event" in Europe'. *Reviews in Palaeobotany and Palynology* 83, 217–26.

9. THE END-TRIASSIC MASS EXTINCTION

— Blackburn, T. J., Olsen, P. E., Bowring, S. A, et al. 2013. 'Zircon U-Pb geochronology links the end-Triassic extinction with the Central Atlantic Magmatic Province'. *Science* 340, 941–45.
— Buckland, W. and Conybeare, W. D. 1824. 'Observations on the South-western coal district of England'. *Transactions of the Geological Society of London, Series 2* 1, 210–316.
— Chapman, A. 2022. *Caves, Coprolites and Catastrophes: The Story of Pioneering Geologist and Fossil-Hunter William Buckland.* SPCK, London.
— Colbert, E. H. 1958. 'Tetrapod extinctions at the end of the Triassic Period'. *Proceedings of the National Academy of Sciences, USA* 44, 973–77.
— Cross, S. R. R., Ivanovski, N., Duffin, C. J., Hildebrandt, C., Parker, A. and Benton, M. J. 2018. 'Microvertebrates from the basal Rhaetian Bone Bed (latest Triassic) at Aust Cliff, S.W. England'. *Proceedings of the Geologists' Association* 129, 635–53.
— Dal Corso, J., Marzoli, A., Tateo, F., et al. 2014. 'The dawn of CAMP volcanism and its bearing on the end-Triassic carbon cycle disruption'. *Journal of the Geological Society* 171, 153–64.

— Marzoli, A., Renne, P. R., Piccirillo, E. M., Ernesto, M., Bellieni, G. and De Min, A. 1999. 'Extensive 200-million-year-old continental flood basalts of the Central Atlantic Magmatic Province'. *Science* 284, 616–18.
— Rampino, M. R. and Stothers, R.B. 1988. 'Flood basalt volcanism during the past 250 million years'. *Science* 241, 663–68.
— Rigo, M., Onoue, T., Tanner, L. H., et al. 2020. 'The Late Triassic Extinction at the Norian/ Rhaetian boundary: biotic evidence and geochemical signature'. *Earth-Science Reviews* 204, 103180.
— Ruhl, M., Hoooolbo, S. P., Al Suwaidi, A., et al. 2020. 'On the onset of Central Atlantic Magmatic Province (CAMP) volcanism and environmental and carbon-cycle change at the Triassic-Jurassic transition (Neuquén Basin, Argentina)'. *Earth-Science Reviews* 208, 103229.
— Suan, G., Föllmi, K. B., Adatte, T., Bomou, B., Spangenberg, J. E. and Van De Schootbrugge, B. 2012. 'Major environmental change and bonebed genesis prior to the Triassic–Jurassic mass extinction'. *Journal of the Geological Society* 169, 191–200.
— Thorne, P. M., Ruta, M. and Benton, M. J. 2011. 'Resetting the evolution of marine reptiles at the Triassic-Jurassic boundary'. *Proceedings of the National Academy of Sciences, USA* 108, 8339–44.
— Whiteside, J. H., Olsen, P. E., Kent, D. V., Fowell, S. J and Et-Touhami, M. 2007. 'Synchrony between the Central Atlantic magmatic province and the Triassic-Jurassic mass-extinction event?'. *Palaeogeography, Palaeoclimatology, Palaeoecology* 244, 345–67.
— Wignall, P. B. 2015. *The Worst of Times: How Life on Earth Survived Eighty Million Years of Extinctions*. Princeton University Press, Princeton.
— Wignall, P. B. and Atkinson, J. W. 2020. 'A two-phase end-Triassic mass extinction'. *Earth-Science Reviews* 208, 103282.

10. THE UNIVERSAL HYPERTHERMAL CRISIS MODEL

— Copp, C. J. T., Taylor, M. A. and Thackray, J. C. 1999. 'Charles Moore (1814–1881), Somerset geologist'. *Proceedings of the Somerset Archaeological and Natural History Society* 140, 1–36.
— Duffin, C. J. 2019. 'Charles Moore and Late Triassic vertebrates: history and reassessment'. *The Geological Curator* 11, 143–60.
— Jenkyns, H. 1988. 'The early Toarcian (Jurassic) anoxic event – stratigraphic, sedimentary, and geochemical evidence'. *American Journal of Science* 288, 101–51.
— Kump, L. E., Pavlov, A. and Arthur, M. A. 2005. 'Massive release of hydrogen sulfide to the surface ocean and atmosphere during intervals of oceanic anoxia'. *Geology* 33, 397–400.
— Self S., Zhao J., Holasek R. E. et al. 1996. 'The atmospheric impact of the 1991 Mount Pinatubo eruption'. In *Fire and Mud: Eruptions and Lahars of Mount Pinatubo, Philippines* (ed. C. G. Newhall and R. S. Punongbayan). Philippine

Institute of Volcanology and Seismology/University of Washington Press, Quezon City/Seattle, pp. 1098–1115.
— Sinha, S., Muscente, A. D., Schiffbauer, J. D, et al. 2021. 'Global controls on phosphatization of fossils during the Toarcian Oceanic Anoxic Event'. *Scientific Reports* 11, 24087.
— Vasseur, R., Lathuiliere, B., Lazar, I., et al. 2021. 'Major coral extinctions during the early Toarcian global warming event'. *Global and Planetary Change* 207, 103647.
— Wignall, P. B. 2015. *The Worst of Times: How Life on Earth Survived Eighty Million Years of Extinctions*. Princeton University Press, Princeton.
— Williams, M., Benton, M. J. and Ross, A. 2015. 'The Strawberry Bank Lagerstätte reveals insights into Early Jurassic life'. *Journal of the Geological Society* 172, 683–92.

11. THE ANGIOSPERM TERRESTRIAL REVOLUTION

— Arnold, E. N. and Poinar, G. 2008. 'A 100 million year old gecko with sophisticated adhesive toe pads, preserved in amber from Myanmar'. *Zootaxa* 1847, 62.
— Benton, M. J., Wilf, P. and Sauquet, H. S. 2022. 'The Angiosperm Terrestrial Revolution and the origins of modern biodiversity'. *New Phytologist* 233, 2017–35.
— Friis E. M., Crane P. R. and Pedersen K. R. 2011. *Early Flowers and Angiosperm Evolution*. Cambridge University Press, Cambridge.
— Friis, E. M. and Skarby, A. 1981. 'Structurally preserved angiosperm flowers from the Upper Cretaceous of southern Sweden'. *Nature* 291, 484–86.
— Ross, A. J. 2018. Burmese (Myanmar) amber taxa, on-line checklist v.2018.2 https://www.nms.ac.uk/media/1158001/burmese-amber-taxa-v2018_2.pdf
— Sokol, J. 2019. 'Fossils in Burmese amber offer an exquisite view of dinosaur times—and an ethical minefield'. *Science*, https://www.science.org/content/article/fossils-burmese-amber-offer-exquisite-view-dinosaur-times-and-ethical-minefield
— Xing, L., McKellar, R. C., Wang, M., et al. 2016. 'Mummified precocial bird wings in mid-Cretaceous Burmese amber'. *Nature Communications* 7, 12089. Xing, L., McKellar, R. C., Xu, X., et al. 2016. 'A feathered dinosaur tail with primitive plumage trapped in mid-Cretaceous amber'. *Current Biology* 26, 3352–60.

12. THE DAY THE DINOSAURS DIED

— Alvarez, L. W., Alvarez, W., Asaro, F. and Michel, H. V. 1980. 'Extraterrestrial cause for the Cretaceous-Tertiary extinction: experimental results and theoretical interpretation'. *Science* 208, 1095–1108.
— Alvarez, W. 1997. *T. rex and the Crater of Doom*. Princeton University Press, Princeton.
— Barrass, C. 2019. 'Does fossil site record dino-killing impact?'. *Science* 364, 10–11.
— Black, R. 2022. *The Last Days of the Dinosaurs*. St Martin's Press, New York.
— Chao, E. C., Shoemaker, E. M. and Madsen, B. M. 1960. 'First natural occurrence of coesite'. *Science* 132(3421), 220–22.
— DePalma, R. A., Smit, J., Burnham, D. A., et al. 2019. 'A seismically induced onshore surge deposit at the KPg boundary, North Dakota'. *Proceedings of the National Academy of Sciences, USA* 116, 8190–99.
— During, M. A. D., Smit, J., Voeten, D. F. A. E., et al. 2022. 'The Mesozoic terminated in boreal spring'. *Nature* 603, 91–94.
— Henehan, M. J., Ridgwell, A., Thomas, E., et al. 2019. 'Rapid ocean acidification and protracted Earth system recovery followed the end-Cretaceous Chicxulub impact'. *Proceedings of the National Academy of Sciences, USA* 116, 22500–4.
— Hildebrand, A. R., Penfield, G. T., Kring, D. A., et al. 1991. 'Chicxulub crater: a possible Cretaceous/Tertiary boundary impact crater on the Yucatan Peninsula, Mexico'. *Geology* 19(9), 867–71.
— Hull, P. M., Bornemann, A., Penman, D. E., et al. 2020. 'On impact and volcanism across the Cretaceous-Paleogene boundary'. *Science* 367, 266–72.
— Morgan, J. V., Bralower, T. J., Brugger, J. and Wünnemann, K. 2022. 'The Chicxulub impact and its environmental consequences'. *Nature Reviews Earth & Environment* 3, 338–54.
— Preston, D. 'The day the dinosaurs died'. *The New Yorker*, 8 April 2019.
— Shoemaker, E. M. and Chao, E. C. 1961. 'New evidence for the impact origin of the Ries Basin, Bavaria, Germany'. *Journal of Geophysical Research* 66(10), 3371–78.

13. RECOVERY AND THE BUILDING OF MODERN ECOSYSTEMS

— Benton, M. J., Wilf, P. and Sauquet, H. S. 2022. 'The Angiosperm Terrestrial Revolution and the origins of modern biodiversity'. *New Phytologist* 233, 2017–35.
— Carvalho, M. R., Jaramillo, C., de la Parra, F., et al. 2021. 'Extinction at the end-Cretaceous and the origin of modern Neotropical rainforests'. *Science* 372, 63–68.
— Herrera, F., Carvalho, M. R., Wing, S. L., Jaramillo, C. and Herendeen, P. S. 2019. 'Middle to late Paleocene Leguminosae fruits and leaves from Colombia'. *Australian Systematic Botany* 32, 385–408.

— Hooker, J. J., Collinson, M. E. and Sille, N. P. 2004. 'Eocene–Oligocene mammalian faunal turnover in the Hampshire Basin, UK; calibration to the global time scale and the major cooling event'. *Journal of the Geological Society* 161, 161–72.
— Hutchinson, D. K., Coxall, H. K., Lunt, D. J., et al. 2021. 'The Eocene-Oligocene transition: a review of marine and terrestrial proxy data, models and model-data comparisons'. *Climate of the Past* 17, 269–315.
— Lyson, T. R., Miller, I. M., Bercovici, A. D., et al. 2021. 'Exceptional continental record of biotic recovery after the Cretaceous-Paleogene mass extinction'. *Science* 366, 977–83.
— McInherney, F. A. and Wing, S. L. 2011. 'The Paleocene-Eocene Thermal Maximum: a perturbation of carbon cycle, climate, and biosphere with implications for the future'. *Annual Review of Earth and Planetary Science* 39, 489–516.
— Simpson, G. G. 1937. 'The Fort Union of the Crazy Mountain Field, Montana, and its Mammalian Fauna'. *Bulletin of the United States National Museum* 169, 1–287.
— Simpson, G. G. 1940. 'The case history of a scientific news story'. *Science* 92, 148–50.
— Stehlin, H. G. 1910. 'Remarques sur les faunules de mammifères des couches éocènes et oligocènes du Bassin de Paris'. *Bulletin de la Société Géologique de France, Série 4* 9, 488–520.

14. COOLING EARTH

— Barnosky, A. D., Koch, P. L., Feranec, R. S., Wing, S. L. and Shabel, A.B. 2004. 'Assessing the causes of Late Pleistocene extinctions on the continents'. *Science* 306, 70–75.
— Coxall, H. K., Wilson, P. A., Pälike, H., Lear, C. H. and Backman, J. 2005. 'Rapid stepwise onset of Antarctic glaciation and deeper calcite compensation in the Pacific Ocean'. *Nature* 433, 53–57.
— Elsworth, G., Galbraith, E., Halverson, G. and Yang, S. 2017. 'Enhanced weathering and CO_2 drawdown caused by latest Eocene strengthening of the Atlantic meridional overturning circulation'. *Nature Geoscience* 10, 213–16.
— Haynes, G. 2022. 'Sites in the Americas with possible or probable evidence for the butchering of proboscideans'. *Paleoamerica* 8, 187–214.
— Lauretano, V., Kennedy-Asser, A. T., Korasidis, V. A., et al. 2021. 'Eocene to Oligocene terrestrial Southern Hemisphere cooling caused by declining pCO_2'. *Nature Geoscience* 14, 659–64.
— Lyle, M., Gibbs, S., Moore, T. G. and Rea, D. K. 2007. 'Late Oligocene initiation of the Antarctic Circumpolar Current: evidence from the South Pacific'. *Geology* 35, 691–94.

— Roberts, A. 2018. *Evolution: The Human Story* (2nd edn). Dorling Kindersley, London.
— Shipman, P. 1986. 'Palaeoecology of Fort Ternan reconsidered'. *Journal of Human Evolution* 15, 193–204.
— Stehlin, H. G. 1910. 'Remarques sur les faunules de Mammifères des couches éocènes et oligocènes du Bassin de Paris'. *Bulletin de la Société géologique de France* (4)9, 488–520.
— Stewart, M., Carleton, W. C. and Groucutt, H. S. 2021. 'Climate change, not human population growth, correlates with Late Quaternary megafauna declines in North America'. *Nature Communications* 12, 965.
— Stuart, A. J. 2021. *Vanished Giants: The Lost World of the Ice Age*. Chicago University Press, Chicago.
— Surovell, T. A., Pelton, S. R., Mackie, M., et al. 2021. 'The La Prele Mammoth Site, Converse County, Wyoming, USA'. In *Human-Elephant Interactions: From Past to Present* (ed. G. E. Konidaris, R. Barkai, V. Tourloukis and K. Harvati). Tübingen University Press, Tübingen, Germany, pp. 303–20.
— Van Couvering, J. A. 1999. 'Book Review: Louis S. B. Leakey: Beyond the Evidence, edited by Martin Pickford'. *International Journal of Primatology* 20, 291–94.
— Werdelin, L. and Sanders, W. J. 2010. *Cenozoic Mammals of Africa*. University of California Press, Berkeley.

15. THE INDUSTRIAL AGE

— Bar-on, Y. M., Phillips, R. and Milo, R. 2018. 'The biomass distribution on Earth'. *Proceedings of the National Academy of Sciences, USA* 115, 6506–11.
— Crutzen, P. J. 2002. 'Geology of mankind – The Anthropocene'. *Nature* 415, 23.
— Ellis, E. C. 2018. *Anthropocene: A Very Short Introduction*. Oxford University Press, Oxford.
— Fuller, E. 2015. *The Passenger Pigeon*. Princeton University Press, Princeton.
— Hume, J. P. 2006. 'The history of the dodo *Raphus cucullatus* and the penguin of Mauritius'. *Historical Biology* 18, 69–93.
— Kitchener, A. C. 1993. 'On the external appearance of the dodo, *Raphus cucullatus* (L, 1758)'. *Archives of Natural History* 20, 279–301.
— MacArthur, R. and Wilson, E. O. 1967. *The Theory of Island Biogeography*. Princeton University Press, Princeton.
— Oreskes, N. and Conway, E. M. 2010. *Merchants of Doubt: How a Handful of Scientists Obscured the Truth on Issues from Tobacco Smoke to Global Warming*. Bloomsbury Press, London.
— Steffen, W., Grinevald, J., Crutzen, P. and McNeil, J. 2011. 'The Anthropocene: conceptual and historical perspectives'. *Philosophical Transactions of the Royal Society A* 369, 842–67.

Acknowledgments

I am grateful to many academic colleagues around the world who answered my questions and provided images for the book. In particular, I thank Chris Duffin (London), Melanie During (Uppsala), David Harper (Durham), Michael Henehan (Potsdam and Bristol), Adrian Lister (London), Alex Liu (Cambridge), Rowan Martindale (Austin, Texas), and Peter Wilf (University Park, Pennsylvania) for reading and commenting on individual chapters.

I thank Ben Hayes at Thames & Hudson for commissioning the book and discussing its shape and form, Joanne Murray for firm but fair copy-editing, Louise Thomas and Celia Falconer for looking after the images, Matt Young for his elegant book design, India Jackson for editorial help, Caitlin Kirkman for promotion and Jen Moore for seeing the book through from start to finish.

Sources of Illustrations

Illustrations are listed first by page and then by plate number.
a = above, **b** = below, **l** = left, **r** = right

2 Richard Bizley / Science Photo Library; **9** Nationaal Archief, The Hague / Archieven van de Compagnieën op Oost-Indië 1.04.01 Inventory Number 136; **15** Keith Chambers / Science Photo Library; **19 l** Sabena Jane Blackbird / Alamy Stock Photo; **19 r** Zeytun Travel Images / Alamy Stock Photo; **24** Ikonya / Alamy Stock Photo; **28** © Michael Böttinger, Deutsches Klimarechenzentrum; **38** Mary Caperton Morton / the Blonde Coyote; **41 a** Smithsonian National Museum of Natural History, Washington, D.C. Photo Michael Brett-Surman (USNM83935); **41 b** FLHC24 / Alamy Stock Photo; **57** Photos courtesy Professor Huang Bing; **59** agefotostock / Alamy Stock Photo; **83** Album / Alamy Stock Photo; **86** Dimitrii Meinikov / Alamy Stock Photo; **95** Simon Colmer / Nature Picture Library; **114** © University of Bristol / Drawing John Sibbick; **120** Photo Michael J. Benton; **122** © Dr. Shixue Hu, China Geological Survey; **125** Drawing Dr. Feixiang Wu; **131** Silvia Anac; **133** The Natural History Museum, London / Alamy Stock Photo; **147** Michael David Murphy / Alamy Stock Photo; **151** Museum de Toulouse. Photo Didier Descouens (MHNT.PAL.CEP.2001.105); **161** Bath Royal Literary and Scientific Institution. Photo Matt Williams (BRLSI M1297); **170** Westend61 / Alamy Stock Photo; **177** from George O. Poinar and Kenton L. Chambers, 'Tropidogyne pentaptera, sp. nov., a new mid-Cretaceous fossil angiosperm flower in Burmese amber',

Index

Page numbers in *italics* refer
to illustrations; page numbers
in Roman numerals refer to
plate illustrations

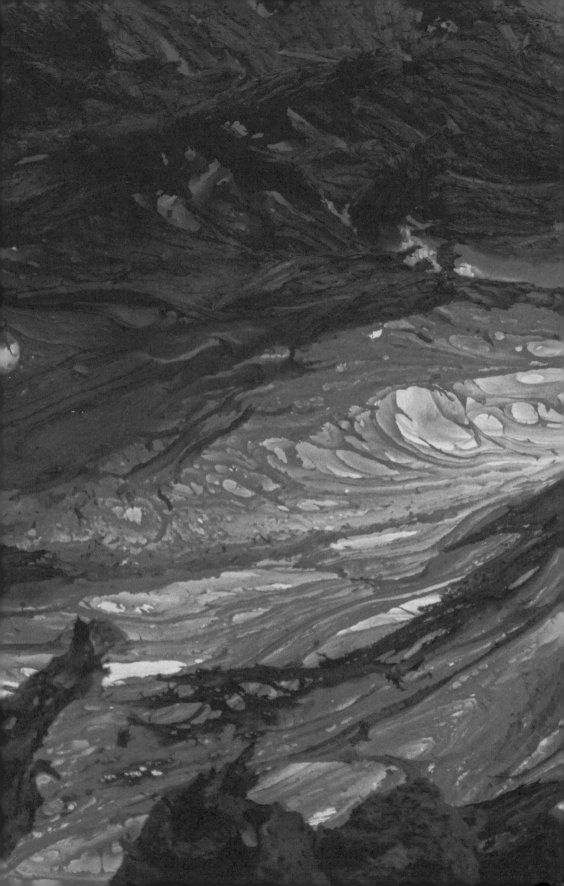